MW01519168

Up-Coast

Forests and Industry on British Columbia's North Coast, 1870–2005

Up-Coast

Forests and Industry

on British Columbia's

North Coast,

1870–2005

Richard A. Rajala

ROYAL **BC** MUSEUM
Victoria, Canada

Published by the Royal BC Museum Corporation, 675 Belleville Street, Victoria, British Columbia, V8W 9W2, Canada.

Cover photographs: (front) Sitka Spruce, Queen Charlotte Islands. BC Archives NA-07210; (back) Whalen Pulp & Paper Mills share certificate, by Kim Martin, © Royal BC Museum.

Maps on pages viii, 149 and 172, and photograph on page 230 © Royal BC Museum. Most of the photographs are from the BC Archives; others, from the City of Vancouver Archives, University of British Columbia (UBC) Library and Vancouver Public Library, are reprinted with permission.

Edited by Andrea Scott and Gerry Truscott.

Designed and typeset by Gerry Truscott, RBCM, in Baskerville 10/12 (body) and GillSans 9/11 (captions).

Cover designed by Chris Tyrrell, RBCM.

Printed in Canada.

Library and Archives Canada Cataloguing in Publication Data

Rajala, Richard Allan, 1953-

    Up-coast : forests and industry on British Columbia's north coast, 1870-2005

    Includes bibliographical references and index.
    ISBN 0-7726-5460-3

    1. Forests and forestry – British Columbia – Pacific Coast – History. 2. Forest products industry – British Columbia – Pacific Coast. 3. Forest management – British Columbia – Pacific Coast. 4. Pacific Coast (B.C.) – History. 5. Pacific Coast (B.C.) – Social conditions. I. Royal BC Museum. II. Title.

SD146.B7R34 2005   338.1'7498'097111   C2005-960237-6

Contents

The publisher gratefully acknowledges the support of the Social Sciences and Humanities Research Council of Canada (SSHRC) and the Natural Sciences and Engineering Research Council of Canada (NSERC), who provided major funds through the Coast Under Stress Project via the SSHRC Major Collaborative Research Initiatives program.

Funding and support for this program was also provided by the host universities – Memorial University of Newfoundland and the University of Victoria – and partners such as the Royal BC Museum Corporation.

For more information visit: www.coastsunderstress.ca

Preface

This book has a long history of its own. It began several years ago with news of a major research initiative on coastal communities by the Coasts Under Stress Project (CUS), headed by Rosemary Ommer. The University of Victoria, where I taught sessionally for the History Department, was to be one of the host institutions. The Royal British Columbia Museum (RBCM) would also become involved, bringing with it a wide range of expertise. Ultimately, Rosemary Ommer invited me to join the CUS research team as a post-doctoral fellow at the RBCM, where I enjoy status as a Research Associate. That gave me the opportunity to participate in CUS's innovative, multi-disciplinary investigation of social, economic and ecological sustainability among Canada's resource-dependent coastal communities. This book represents a small part of a much larger collaborative effort, then, one that has seen scholars on both coasts join forces under the CUS umbrella in the search for ways to achieve a sustainable human relationships with coastal ecosystems in a global economy.

I am very grateful to several individuals and institutions for these opportunities. Funding from the Social Sciences and Humanities Research Council of Canada and the Natural Science and Engineering Research Council of Canada made the entire CUS endeavour possible. CUS Project Director Rosemary Ommer approved my participation. John Lutz helped connect historians with the CUS project. At the RBCM, Bob Griffin and Lorne Hammond provided leadership, advice and a good deal of support throughout the long research and writing phases. Particular thanks goes to Lorne for handling administrative tasks and ordering the images. RBCM editor Gerry Truscott managed the publication process with skill and flexibility. Discussion of Ocean Falls builds on an earlier project for the Canadian Museum of Civilization, initiated by Dan Gallacher. Finally, this book, like the others, would never have seen the light of day without the participation of my wife, Jean. This book is dedicated to her.

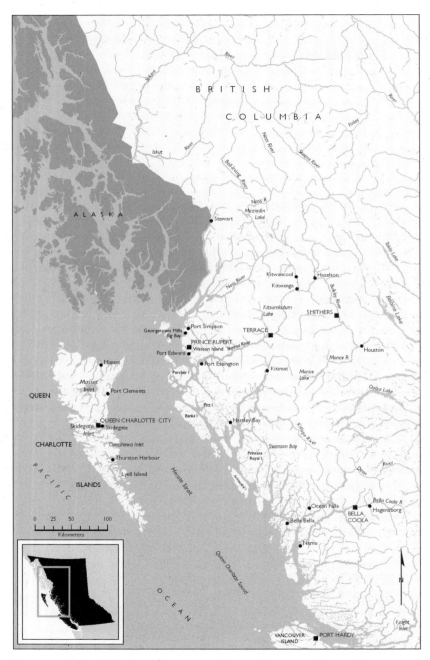

Up-coast British Columbia: major communities and features of the north coast.

Introduction

This account seeks to shed light on the relationships between forests, industry, people and communities along the British Columbia coast from Seymour Inlet to the Cassiar district. It takes in the Queen Charlotte Islands, and extends east to consider the Skeena and Nass river valleys. The region as a whole has received little attention from students of British Columbia forest history, although in recent decades scholars from a variety of disciplines have produced a good deal of work on First Nations issues and modern environmental politics. I hope to contribute by integrating these themes within a more broadly based social and environmental history of industrialization, community development and government policy making, culminating in an account of the region's recent experience of industry contraction, land-use conflict, and First Nations claims to traditional territories.

In discussing the period from roughly 1870 to early 2005 I have crossed the boundary that typically separates history from political science or journalism. But as recent events unfolded in media coverage of First Nations land claims, forest and environmental policy, and deindustrialization, the urge to include such topics proved irresistible. The history of the Prince Rupert pulp mill, for example, seemed more vivid in light of recent developments, and those developments are better understood in terms of the mill's historic significance in the region's postwar economy. The same logic can be applied to the Nisga'a settlement, which comes after decades of conflict over forest practices or the Tree Farm Licence associated with pulp manufacturing at Prince Rupert.

This story begins in the late 19th century, however, when industrial capitalism began penetrating aboriginal space in the form of salmon canneries and rudimentary sawmills. In addition to cutting box and construction material for the canneries, these water-powered plants provided lumber for dwellings, docks and bridges at mission settlements, aboriginal villages, and the Norwegian colony of Bella Coola. By the turn of the century many aboriginal males had worked handlogging and fishing into their seasonal round of subsistence and commercial activities.

1

The wood products market remained little more than a local, sporadic, and minor supplement to fishing, canning and trapping, however, until the provincial state opened the hemlock, spruce and cedar forests to mass production by offering huge pulp leases to major investors in 1901. Attractive terms lured British, Canadian and American capital to the coast, eventually creating a foundation for developments at Ocean Falls and Swanson Bay. Not satisfied with that promotional effort, Premier Richard McBride unleashed a speculative mania by sweetening the deal on sawmill quality stands a couple of years later. The offer of secure tenure produced an enormous infusion of capital between 1905 and 1907, resulting in the staking of much of the coast's best forestland.

The legacy of these policies remains with us to this day, but immediate development north of Seymour Inlet proved slow in coming. Aboriginal handloggers found their access to tidewater timber constrained by the flood of new licences and leases. It wasn't until a 1909 ownership change, three years after government approval of the 80,000-acre pulp lease, that development got underway at Ocean Falls. Another three years passed before pulp production began there, but by the end of 1912 debt had claimed the enterprise. Swanson Bay's early history was even more tortured, producing a dizzying flurry of corporate reorganizations, ownership shifts, and aborted start-ups. By 1913, after a few bursts of activity, the mill and town site at Swanson Bay was silent as well. Handloggers, initially enthusiastic about the prospect of healthy log markets, went back to supplying the few, small sawmills as the 1913 recession set in.

Along the Skeena River, the new century offered hope for a resource-based economy linked to North American markets by the Grand Trunk Pacific Railway (GTP). Timber cruisers in the employ of outside investors combed the Skeena watershed, staking claim to commercial stands, and the laying of rails east from the new town of Prince Rupert created a tie industry by 1907. Small mills appeared at Prince Rupert, Terrace and other points, cutting construction lumber for town site development and settler homes. The boom faded quickly here too, however, as Grand Trunk construction slowed despite government subsidies.

No area reflected the direction of future economic trends more distinctly than the Queen Charlotte Islands, where a few large syndicates locked up millions of feet of magnificent Sitka Spruce between 1908 and 1912. A couple of small sawmills even went up in conjunction with their arrival, amid boasts of huge lumbering and paper developments to come. Those promises would, in fact, go unfilled in both the short and long-term. The investors retained possession of their timber rights by paying paltry annual rentals to the Crown in anticipation of rising values, and World War I would bring the first windfall profits.

Elsewhere along the coast the operators of small mills at Georgetown, Big Bay, Port Essington, Hartley Bay, Namu and Bella Coola, along with a few others scattered along the Skeena River from Prince Rupert to Terrace,

comprised the roster of a fledgling industry that awaited the development
of a maritime or rail trade to distant markets. Handloggers, operating without the benefit of sophisticated steam technology that had already transformed south-coast logging into a factory-like mass-production regime, cut
tidewater timber on a seasonal basis. Horses and oxen drew logs short distances to some plants, and aboriginal labour figured prominently in harvesting and processing the resource. Forest capitalism came to the central
and north coast not in the form of a sudden force overturning everything
in its path, then, but in a tentative series of waves. But one fundamental
shift had occurred: accessible forests had been seized as a form of private
property, and this process of corporate domination would continue to
unfold despite First Nations defence of their title to traditional territory.

World War I brought both dramatic change and structural entrench-
ment to the up-coast forest economy. The GTP's completion in 1914
cheered boosters, but the war brought Prince Rupert's boom to an abrupt
end, and it would not recover until the next world conflict. The railway
itself was in poor health, although lumbermen held out hope that low rates
would allow their products entry into the prairie market. The Panama
Canal offered a potential outlet to the eastern seaboard, but neither of these
options stimulated exports during the early war years. As immigration to
the region dried up and enlistments drained the local population, settlement along the Skeena slowed. Only the brisk trade in salmon and halibut
box material and the railway's ongoing demand for ties and pilings offset
the dismal outlook.

Nor were up-coast residents wholeheartedly in favour of the forest
economy's emerging structure. Handloggers endured interminable delays
in Victoria's processing of licence applications, fuelling discontent at a timber allocation system that favoured speculation by outside corporate interests. But that expression of hinterland grievance could not approach the
frustration of up-coast First Nations. By the time of the 1913 McKenna-
McBride Royal Commission on reserve allocations, Ottawa had given up
any pretence of pressuring the province to extinguish aboriginal title
through treaty. The hearings did provide a forum for First Nations to
demand greater access to the forest resource, but the commissioners denied
requests for reserve extensions if cutting rights had been awarded on the
lands in question.

Handlogging opportunities did pick up after 1914, however, thanks in
large part to the Crown Willamette Paper Company's acquisition of the
Ocean Falls facility, town site and timber holdings. Pacific Mills, a Crown
subsidiary, initiated newsprint production in 1917, going on to develop one
of the province's most renowned company towns. By 1918, Ocean Falls
had perhaps 2,000 residents, housed separately according to race, but all in
close proximity to the mill along the cramped shore of Cousins Inlet. Swanson Bay also got a temporary new life when Ontario lumberman James
Whalen folded that operation into an ambitious coast-wide structure that

included pulp plants at Port Alice and Howe Sound. Whalen's scheme for control over the province's pulp industry would quickly falter, but for a brief time prospects seemed bright for the emergence of a competitive up-coast log market.

Exceeding all these developments in sparking short-term up-coast forest industrialization was the brief war-inspired Sitka Spruce boom on the Queen Charlotte Islands. Imperial Munitions Board demand for aircraft components unleashed an onslaught on the previously ignored stands bordering Masset Inlet, Cumshewa Inlet, Skidegate Inlet, Lyell Island and other points. Sophisticated cable logging systems driven by massive steam engines arrived, along with hundreds of loggers, to feed a suddenly limitless market that offered the added bonus of guaranteed profits. The spruce drive even resulted in the construction of several sawmills, but most logs left the islands in rafts bound for Ocean Falls and Swanson Bay. After the armistice, most of those few mills vanished, as did the vast majority of the loggers, leaving behind enormous waste sanctioned by the government-industry partnership's sacrifice of conservation to victory and profit. With little left in the way of an industrial infrastructure, the islands slipped into their role as a supplier of raw material for mainland mills to the south, one the area would never escape.

The up-coast forest industry experienced the full effects of a sharp post-war recession, and the lumber sector remained stagnant for much of the 1920s. Troubled by high freight rates, the inability to develop a consistent waterborne trade with Asian markets, and stands that offered a low proportion of construction-grade timber, sawmillers operated sporadically on narrow profit margins. Even J.S. Emerson's modern Prince Rupert plant, built at the height of the spruce boom, limped along before burning to the ground in a 1925 fire. Only the highest grades of clear Sitka Spruce lumber promised a profit in the rail market, and the mainland offered such timber only in isolated stands. Even prime Queen Charlotte Islands spruce towed to Prince Rupert for sawing could not overcome the freight rate advantage Canadian National Railway officials afforded Prince George operators after Ottawa's takeover of the GTP.

A slight upturn in 1923 prompted H.R. MacMillan, part-owner of the Georgetown mill at Big Bay, to head up efforts to develop a market for hemlock in Japan. The larger CNR operators between Prince Rupert and Terrace participated, but falling prices doomed the initiative within a year. That left Olaf Hanson's tie industry and a new sector devoted to the cutting of cedar telephone and telegraphic poles, to take up the slack. Blessed with timber suitable to the cutting of both products, the Skeena River valley became home to a diverse industry that resembled the agri-forestry model of eastern Canadian lumbering regions. Homesteaders cut ties and poles from their own pre-emptions or worked in one of Hanson's camps during the winter, accumulating cash for "stumpfarm" development. Wage labour in logging, sawmilling, mining and fishing offered other seasonal alternatives. Such flexibility might confer both a measure of independence from any single employer and a means of coping with seasonal and cyclical economic fluctuations.

The wisdom of sectoral diversity received reinforcement in 1923, when Swanson Bay disappeared from the up-coast industrial landscape. Rescued from the 1919 collapse of the entire Whalen organization by new backers, the operation succumbed to a host of factors. The entire physical plant had been allowed to deteriorate along with the town site, the 1923 Tokyo earthquake destabilized Whalen's chief pulp market, and the low quality of Swanson Bay timber hindered the profitability of both pulp and lumber operations. After acquiring Whalen's assets in 1925, the new B.C. Pulp and Paper Company decided to rescue the Woodfibre and Port Alice mills, and let the up-coast facility die. Only a few handloggers and a caretaker continued to live there, the latter protecting the original pulp leases from encroachment.

While brush reclaimed Swanson Bay, Pacific Mills developed Ocean Falls into the hub of the mid-coast forest industry. Indeed, the firm controlled both its town and the entire log market from Bella Coola to the Queen Charlotte Islands. At Ocean Falls the provision of housing, schools and other urban amenities offered resource workers the opportunity to keep families intact, and in conditions that surpassed those of typical camps and mill towns. But corporate paternalism involved both carrot and stick. In this era, firms like Pacific Mills tolerated no challenge to their ownership rights over land, housing, mills, store, utilities and the all-important dock, the sole point of entry to the community.

If Pacific Mills held the upper hand with its workforce, it proved equally well positioned to wring concessions from government. As the largest employer by far on the central coast, in addition to owning and operating the region's largest settlement, executives typically found a receptive audience in Victoria. During the 1920s Pacific Mills expanded its mainland timber rights by well over a billion feet, purchased rights to millions more on the Queen Charlottes, and participated in the timber sale program. As early as 1922, District Forester E.C. Manning warned that short-sighted management of the firm's pulp leases would have eventual consequences.

While operating company camps from time to time as conditions warranted, Pacific Mills drew most of its log supply from contractors and handloggers. Even while Swanson Bay operated, the two operations avoided competing for fibre, but the northern mill's demise eliminated even a semblance of a competitive log market. Notorious for driving hard bargains with its suppliers, the firm exploited its control over both the resource and the market to drive down raw material costs. Reliant upon Pacific Mills for advances at the start of the logging season and log purchases at its end, contractors resembled southern sharecroppers locked into a cycle of dependence that left little room for the emergence of a truly independent logging sector.

Pacific Mills' quasi-monopoly and ruthless drive to lower fibre costs contributed to a general disregard of silvicultural and utilization standards. The Forest Branch had neither the staff nor the scientific data to support rigid regulation of the hard-pressed contractors, even if agency personnel possessed the will. E.C. Manning issued gloomy forecasts of reproduction on clearcuts along the coast during the early 1920s, and his successor Parker Bonney was equally pessimistic. Not until 1927 did the Forest Branch initiate regeneration surveys north of Seymour Inlet, an area where the damp climate eased anxiety over the danger of fire destroying natural seedling growth on cutovers.

Of more immediate concern to foresters was the failure of up-coast lumbering to enjoy the benefits of a late-1920s boom in demand for provincial forest products. Regional forests simply did not produce a volume of high-grade lumber commensurate with the Douglas-fir zone to the south, and high rail freight rates made marketing of the lower grades difficult. Lower Mainland and Vancouver Island mills also enjoyed a cost advantage in waterborne rates to eastern markets. The Queen Charlotte Islands held an abundant supply of high-grade spruce, but Pacific Mills and the Powell River Company had the financial resources to lock up much of the accessible timber not already held under lease, licence or Crown grant. Islanders expressed disapproval of government decisions that seemed to sacrifice local economic development to the demands of distant holding companies and mainland paper producers, but policy makers in the provincial capital of Victoria consistently favoured their corporate clients.

All sectors of the up-coast forest industry felt the impact of the Great Depression, sawmilling less sharply than others simply because it had less distance to fall. Operators, competing for the dwindling orders from the canneries, slashed wages and cut timber sporadically, even accepting logs and farm produce in exchange for lumber. Tie and pole demand practically vanished during the early 1930s, leaving settlers with little alternative but to concentrate on homestead improvement or accept relief. Cooperative societies sprang up to enable settlers to handle CNR tie orders directly, an attempt to eliminate the contractors' commission charge and maintain a living wage. Pacific Mills managed to ride out the Depression by cutting wages and making life even more difficult for handloggers and contractors. The firm even engaged Managerial Consultant Charles Bedaux to conduct time studies as a basis for a bonus payment plan that made mill-employee earnings contingent upon productivity increases.

Clearcutting standards on the Charlottes declined even further under the weight of Depression-era market constraints. T.A. Kelley, A.P. Allison and J.R. Morgan – the main contractors for Pacific Mills and the Powell River Company – concentrated on the profitable Sitka Spruce, taking no care to protect residual hemlock and cedar timber from damage during high-speed cable yarding operations. The introduction of logging trucks, crawler tractors, and even a railway to penetrate Powell River Company holdings at Cumshewa Inlet reflected the diminishing availability of tidewater timber on the Charlottes by the end of the decade.

While foresters began raising mild alarms about the unsatisfactory level of regeneration on larger clearcuts, they were perhaps more concerned about a pace of exploitation that fell far below the region's sustained-yield capacity. Another pulp mill would help remedy that situation, but repeated rumours of a Prince Rupert plant only created a cycle of public enthusiasm and cynicism. A growing number of south-coast operators had more concrete plans for northwestern forests. Having depleted their own holdings, and with little room for relocation in the Vancouver Forest District, some began moving up-coast. The result would be a "drain" of resources to Lower Mainland mills, facilitated by efficient log-rafting technologies that made it more cost-effective to tow the cut south than to expand the northwest's manufacturing base.

The Queen Charlotte Islands represented the clearest example of hinterland resources being drawn off without appreciable local benefit. World War II only exacerbated that trend, unleashing another spruce drive that again exposed the provincial government's woeful lack of investment in silvicultural science. Market recovery had brought the annual spruce harvest to several times its sustained-yield capacity on South Moresby Island by 1940, but the Forest Branch had not yet conducted a systematic study of how best to promote natural reproduction of the species. Having focused its limited research budget on the problem of renewing south-coast Douglas-fir forests, the agency did little to assess the relationship of logging practice to forest regeneration in the northwest. What insight foresters had gained did not inspire confidence.

Production and profit would trump conservation during the World War II as it had in World War I, despite a near monopoly in Commonwealth markets and the 1942 establishment of Aero Timber Products, a Crown corporation to oversee airplane spruce output. Some foresters thought that tractor logging of Sitka Spruce on a selective basis offered a potential solution, but wartime proved no more conducive to rigorous silvicultural study than the Depression. And Aero turned out to be anything but a silvicultural saviour, adapting overhead cable systems to selective logging in pursuit of maximum spruce production. The other Queen Charlotte operators followed suit, ravaging mixed forests to draw out select logs suitable for manufacture into aircraft components.

Wartime labour shortages produced greater opportunities for First Nations in the Skeena valley sawmills, which enjoyed a vigorous local market due to the construction of defence infrastructures at Prince Rupert and Terrace. Profits might even be found in the rail trade to prairie and eastern markets. Aboriginal males played the labour market to advantage, abandoning pole cutting when rising fish prices promised higher earnings. Pacific Mills scoured the country for labour, eventually staffing its sawmill entirely with women. But labour's rising fortunes did not come without struggle, as witnessed on the Queen Charlottes when the province's spruce drive placed loggers in a position of advantage. Led by the International Woodworkers of America's "Loggers' Navy", the spruce loggers played a pivotal role in the union's achievement of collective bargaining rights in coastal camps and mills by the end of the war.

In the immediate postwar decades, the conjunction of a booming western industrial economy, timber scarcity to the south and the adoption of sustained-yield policies signalled the up-coast forest economy's long-awaited maturity. An accommodating freight rate structure and healthy continental construction markets produced a tremendous expansion in sawmill capacity along the Skeena River, triggering competition for the limited amount of accessible timber. Pacific Mills boosted production at Ocean Falls, drawing logs from company camps on the Queen Charlottes, northern Vancouver Island and mainland points, in addition to contractors and handloggers. Alaska Pine and Cellulose knitted the old Swanson Bay pulp leases and Queen Charlotte holdings into a coast-wide fibre network to feed southern pulp plants and sawmills.

But in one respect the postwar structure of corporate control marked no departure from the prewar pattern that siphoned off northwestern resources to the benefit of communities elsewhere. In fact, intense demand from south-coast manufacturing centres only increased southbound log traffic. Prince Rupert Forest District officials made tentative appeals for the reservation of timber for up-coast industrial development. Their recommendations went unheeded, but the provincial government's 1947 sustained-yield legislation at long last brought the Prince Rupert pulp mill from the realm of myth to reality.

The device chosen to achieve this transformation was the Tree Farm Licence (TFL), conceived as an industry-government partnership in forest management. The brainchild of Chief Forester C.D. Orchard, the tenure responded to timber capital's need for assured access to raw material in developing a new generation of pulp-and-paper plants around the province. Orchard's abundant faith in private enterprise, coupled with a desire to promote investment and community stability, inspired his creation. Licences would relieve his agency of the responsibility for managing the tenures, pursue their own private interest in managing their holdings on a profitable crop-rotation basis, and thereby fulfil the public's interest in responsible stewardship. Or so the theory held.

The Celanese Corporation's Tree Farm Licence No. 1, initially taking in over 668,000 acres of Crown forestland in the watersheds of the Nass and Skeena rivers, was the new tenure system's flagship. Through its Columbia Cellulose (Colcel) subsidiary the firm promised to revitalize the up-coast economy, rescue Prince Rupert from its decades-long economic doldrums by building its $25 million Watson Island pulp mill, and make wise-use of a resource for the general welfare. Understandably, many rejoiced at hearing such confident predictions, but others had reservations about the province's largesse that with the stroke of a pen awarded so much control over the region's economic destiny to a single corporate entity. Their concerns proved more valid than the boosters' enthusiasm.

Colcel encountered problems from the outset, failing in its attempt to drive logs down the Skeena River from its most accessible holdings. Forest Service officials reported friction over Colcel management policies. Highly mechanized logging operations limited employment gains, and took a heavy toll on fragile forest ecosystems. Nor were the economic impacts uniformly positive. Skeena Valley sawmillers competed fiercely for remaining timber, and long-held hopes that another pulp mill would finally reward up-coast contractors with a competitive log market proved groundless. Another wave of industry consolidation brought new multinationals to dominance in the 1950s. Crown Zellerbach folded Ocean Falls into a vast structure that extended from Vancouver to the Queen Charlotte Islands. Rayonier took over Alaska Pine, bringing a new 300,000-acre South Moresby Island TFL into its fibre-production system. A 1960 merger of H.R. MacMillan's interests with the Powell River Company took that organization to the peak of industry integration. New self-dumping log barges provided these firms with an even more efficient method of moving up-coast logs to southern mills, helping turn the northerly trickle of operators into a flood by the mid 1950s.

Inevitably, the scale of postwar industrialization provoked protest from First Nations. The new corporate order not only wreaked havoc upon hunting, gathering and trapping territories, but also closed off aboriginal entrepreneurial access to timber. The Gitxsan in the Kitwanga-Kitwancool appear to have been the most determined opponents, engaging in a dispute with the Forest Service throughout the early 1950s. Lacking the capital to compete successfully in the agency's competitive timber sale program, the Gitxsan petitioned for a forest reserve. Citing regulations that mandated public competition for Crown timber, the Forest Service denied all such requests. Major corporations might avoid the rigours of competition through the TFL scheme, then, but First Nations and small operators would gain no such relief.

Some local Forest Service officials were more sympathetic, citing a high illiteracy rate that rendered First Nations' understanding of complex timber-sale procedures difficult. Minister of Lands E.T. Kenney refused to budge, however, maintaining throughout that First Nations were perfectly free to apply for Crown timber, but would receive no special consideration. The government was even less receptive to the more fundamental Gitxsan position that their title to the land itself remained intact, the premise of a 1955 petition demanding a cessation of all timber sales and cancellation of Colcel's Tree Farm Licence pending settlement of the land question.

Gitxsan, Tsimshian and Nisga'a workers occupied a central place in the up-coast forest economy. Many of the latter worked as loggers for Colcel on the tenure that occupied a third of the Nisga'a's traditional lands. As

Frank Calder revitalized their land-claim fight during the early 1960s, Colcel pressed north to the Nass River after depleting its Skeena holdings. Despite an aggressive style of clearcutting, Colcel and the Forest Service were content to let nature take its course in providing a second crop. By the end of 1961 Colcel had planted only about a thousand acres, though it clearcut three times that area in a typical year. Not until the late 1950s did the Forest Service establish a small nursery to meet regional seedling needs, and the agency still lacked the scientific legitimacy needed to introduce clearcutting regulations. Corporate foresters resisted any restrictions on production efficiency when pressed to reduce the size of cutblocks to promote natural reseeding.

By 1960, overtaxed Prince Rupert District foresters faced a flood of TFL applications from firms fleeing the increasingly denuded south coast. Alaska Pine and MacMillan Bloedel would prove victorious at this stage of the timber scramble, both gaining approval of licences that had little geographical logic. Each would mix Vancouver Island blocks with those on the mainland and Queen Charlotte Islands, further entrenching the latter area's status as fibre supplier for the south. District staff continued to question this policy, and went further to express concerns about the north coast's capacity to accommodate all the newcomers. Small operators, ranchers and farmers also had reservations about the locking up of northern resources, using a stronger language of protest.

Like the Queen Charlottes, the Bella Coola River valley represented in microcosm the consequences of multinational forestry along the coast. There, Crown Zellerbach's extensive early 20th-century licences provided raw material for the aging Ocean Falls plant and a new Campbell River paper facility. Residents, some of them members of a small local logging and milling sector, criticized Crown Zellerbach's practices as part of a larger critique of corporate forestry. But Bella Coola's Northcop Logging Company would confront the contradictions of a policy that awarded monopolistic control to multinational companies while leaving smaller interests to weather the storms of a competitive marketplace. Northcop's timber needs ranked second to those of Crown Zellerbach, and Bella Coola would become a site of large-scale corporate logging that generated little or no local manufacturing employment.

The presence of a new pulp mill at Prince Rupert, fed from Colcel's Terrace logging headquarters, seemed to warrant a more positive outlook. Along the Skeena River, sustained-yield policy had created a dynamic force for employment and northern development. Seeing limitless potential for similar modernization, Kenney's successor, Ray Williston increased Colcel's holdings in 1964, awarding the firm a second Nass River Tree Farm Licence to support a new kraft mill at Prince Rupert. Williston also carved out a tenure for Ben Ginter and a Finnish consortium, who won out over Crown Zellerbach and MacMillan Bloedel in a battle for the rights to establish a Kitimat paper facility. That licence, the 41st in the province, extended to Ootsa Lake in the interior, where independents were scrambling to protect their allowable cuts. Williston went on to award timber rights for two new Prince George pulp mills, slicing up the northern timber pie in ways that astounded some observers.

The new mills met the immediate goals of generating revenue, jobs and regional infrastructure development, but new technologies constrained employment benefits. Tractor-mounted shears, the introduction of hydraulic grapples to rubber-tired skidders and overhead yarding systems, and the development of feller-buncher technology stripped labour from logging sites as they boosted productivity. Such innovations were crucial to corporate balance sheets, given the cost of moving raw material over extensive road and water transportation systems. By 1967, Colcel had pushed its main logging road almost to Meziadin Lake on the upper Nass River, utilizing the river for log drives to Portland Inlet. As fibre costs and losses mounted in the 1960s, Colcel gobbled up sawmills to acquire their Skeena valley timber rights. Pollution problems at the old Watson Island sulphite mill added to the firm's woes, as the annual cut rose steadily under loose Forest Service administration of sustained-yield guidelines.

Despite the warning signs, British Columbia's postwar forestry boom maintained its momentum through the 1960s. Social Credit policies ensured industry's secure access to timber at reasonable prices, and the corporate partners fulfilled their part of the bargain by investing in plant expansion. Policy making had settled into a closed process of negotiation involving industry, government and union elites who bickered about how the financial pie would be shared, but closed the door to dissident outsiders. Their world was about to change dramatically, however, as environmentalists, First Nations and community activists arose to challenge the legitimacy of a sustained-yield accord that seemed bent on sustaining capital accumulation at the expense of human and ecological values.

Dave Barrett's New Democratic Party (NDP) government had to confront both the legacy of unrestrained forest exploitation and the winding down of the postwar boom during the early 1970s. Crown Zellerbach closed Ocean Falls, preferring to invest in the modern Elk Falls facility rather than maintain its central coast community. The Celanese Corporation made the same decision, opting to withdraw from its sustained-yield partnership with the province after depleting the Nass Valley forests. The NDP took over the Ocean Falls and Prince Rupert mills, among others, in a pragmatic effort to protect a regional economy undergoing a simultaneous contraction of commercial fishing and fish processing. Up-coast aboriginal bands and communities responded to the looming crisis by pressing for community- and village-based forest tenures. These goals would have been more easily reached had the outgoing Social Credit party deprived Crown Zellerbach of its pulp leases rather than allowing the timber to be routed to Elk Falls.

Disenchantment with multinational forestry mounted along the coast during the early 1970s. The Haida pressed their claim to the Queen Charlotte Islands, as the Barrett government reacted to environmental concerns by imposing tighter regulations on clearcutting practices. All sectors of the industry took the opportunity presented by the Pearse Royal Commission to attack the regulations that threatened private enterprise autonomy. But Pearse's hearings also allowed for presentation of alternative models. The Skidegate Band Council, Nass Valley Communities Association and Kispiox Valley Community Association questioned every aspect of corporate forestry: its lack of concern for fisheries and wildlife, the paucity of regional benefits, and the absence of meaningful local input on policies that left communities reeling with the consequences of environmental degradation, forest depletion and economic decline.

Pearse presented his report to Bill Bennett's Social Credit government in 1978, and while the new Forest Act that followed included a commitment to multiple-use principles, it did little to reverse the process that had seen just 10 firms seize control of 59 per cent of the public forests. The government folded the Canadian Cellulose mill at Prince Rupert into their B.C. Resources Investment Corporation, but extricated themselves from Ocean Falls by shutting the operation down permanently in 1980. A shortage of accessible timber along the central coast contributed to the demise of Ocean Falls, and as the world economy plunged into recession in the early 1980s many areas confronted the legacy of decades of overcutting. Most of the postwar giants withdrew for greener investment pastures, leaving aging mills and stunned communities behind. There were no takers for the Prince Rupert plant, where high wood costs and pollution problems contributed to a gloomy long-term outlook.

As the recession deepened, the fabric of postwar forest policy began to tear from the strain of land-use conflict. On the Queen Charlotte Islands the threat steep-slope logging posed to salmon habitat pitted federal fisheries officials against companies, the province and the International Woodworkers of America. Even more contentious was the logging of Western Forest Products' TFL No. 24 on South Moresby Island, which saw the Haida and environmentalists unite in campaigning for the creation of Gwaii Haanas National Park. That achievement did not detract from Haida determination to secure recognition of their claim to the entire archipelago, and on the mainland the Nisga'a and Gitxsan peoples pursued similar goals.

Gitxsan road blockades during the late 1980s reflected a widespread sense of urgency among First Nations confronted by the province's refusal to acknowledge the existence of aboriginal land title. Gitxsan leaders asked: Would any forests be left standing when their court case reached a conclusion? The early 1991 decision rendered by B.C. Chief Justice Alan McEachern on the *Delgamuukw* case, denying any aboriginal rights to land, unleashed another series of blockades. A more favourable 1993 appeals court ruling led to yet another appeal, and the Supreme Court of Canada's 1997 *Delgamuukw* decision legitimating the concept of aboriginal title. During the same period, Nisga'a leaders pressed for reforms on the Nass River valley Tree Farm Licence that supported the Prince Rupert pulp complex, calling for a halt to logging while treaty talks progressed.

But First Nations were not alone in expressing hostility to the existing structure of decision making in distant corporate and government offices. The village of Hazelton's 1990 Forest Industry Charter of Rights demanded a complete overhaul of a forest policy that placed communities at the mercy of outside corporate entities that sacrificed economic and ecological sustainability for short-term-profit objectives. Nowhere was this tendency more apparent than at Prince Rupert, where Repap Enterprises of Montreal had briefly assumed the role of saviour by purchasing the pulp mill and timber rights from the British Columbia Resources and Investment Corporation's Westar Timber Division in 1986. Repap went on to rack up huge losses, developing acrimonious relations with workers and contractors, while buying up smaller north-coast operations and their timber rights.

By 1998 Repap was ready to walk away from its debt-ridden Skeena Cellulose operation, leaving contractors and suppliers throughout the region with millions of dollars in unpaid bills. Closure of the pulp mill and sawmills in communities as far east as Smithers left over 2,000 unemployed, ample evidence of the regional economy's vulnerability. A last-minute compromise between the New Democratic Party government, creditors and workers saved the enterprise, which operated under public ownership until Gordon Campbell's newly elected Liberal government unloaded it at a bargain basement price to a pair of former Repap executives in 2002. Aggressive in their pursuit of concessions from labour and the province, the new owners failed to attract the capital needed to revive the operation.

While Skeena River communities struggled to cope with the unravelling of Skeena Cellulose's empire, decades of overcutting on the central coast came back to haunt the region Bella Coola came to the same conclusion reached by Hazelton, pleading for a community-controlled Forest Licence to counterbalance the economic and environmental abuses of corporate forestry. Scrambling to reduce harvest rates to sustainable levels, the Ministry of Forests met further pressure on the land-base from environmentalists and First Nations. West Fraser Timber voluntarily relinquished its cutting rights to the Kitlope watershed in 1994, but on the Queen Charlotte Islands a similar scenario of harvest adjustment and Haida pressure unfolded without corporate concessions. Here the *Delgamuukw* decision provided the basis for legal challenges to the Crown's authority to renew MacMillan Bloedel's Tree Farm Licence in 1997 without consultation. The Haida were back in court in 2000, this time in protest of the tenure's sale to Weyerhaeuser.

By that time the dust had begun to settle from the landmark 1998 Nisga'a Agreement, which saw the Crown relinquish ownership of nearly 2,000 square kilometres of Nass watershed land. Despite several provisions designed to ease the concerns of existing rights holders, for the first time an up-coast First Nation could look forward to undisputed forest management authority on traditional territory. Industry organizations grumbled at the treaty's terms, more perhaps out of concern for the precedent than the lost timber values. The fight over what has come to be known as the Great Bear Rainforest raised the stakes for some of the largest coastal firms. Although past logging had taken most of the best and most accessible timber, the remnants were well worth fighting for. Complicating matters were a diverse set of environmental organizations, First Nations claims, the ominous threat of an international boycott, changes in government, and direct action by woodworkers to protect jobs.

One almost needed a scorecard to keep track of the shifting strategies, negotiations and alliances that emerged over the 1993–2003 period. The process would see a premier declare environmentalists enemies of the province, the International Woodworkers of America blockade Greenpeace vessels in the Port of Vancouver, and major companies forge an uneasy alliance with "green" organizations in search of a consensus, an effort resisted by small operators suspicious of a sell-out of their interests. The one constant was an unblinking determination by up-coast First Nations to uphold their sovereignty. The 2001 Great Bear Rainforest Agreement achieved what seemed an unlikely balancing act. Protected area status for the most prized mid-coast watersheds and a shift to ecosystem-based forest practice would defuse the threat of boycotts by customers who had been convinced of the evils of clearcutting. Loggers and smaller operators would be compensated for the withdrawal of rights to the "working forest". Several First Nations accepted a more significant voice in the government's ongoing Central Coast Land and Resource Management Planning process, and access to timber through interim agreements.

For a brief moment, then, peace had broken out in the province's forest wars. But the euphoria gradually turned to resentment when surveys revealed that industry cutting practices had undergone no transformation. These findings, in conjunction with Liberal government cutbacks and splitting of the Ministry of Environment, have caused doubts to surface about the capacity of Gordon Campbell's neo-conservative administration to balance ecological and business welfare. Several mid-coast First Nations received timber rights under recent agreements with the province, suggesting a willingness to accommodate their interests and minimize land-claims protests against logging while treaty talks drag on. What the future holds for the up-coast forest economy is unclear. Will a new structure emerge out of the ruins of the old, one more attuned to the needs of coastal communities? Can an industry historically geared to supplying international commodity markets take on a new, more diversified and sustainable outlook? Can aboriginal and non-aboriginal residents find a way to share equitably in the range of benefits both the old and new forest offers? If history is any guide, the obstacles are enormous.

But neither should history be considered irrelevant in the seeking of solutions. Too often in the past, governments have made policy based on the naïve assumption that healthy corporations would inevitably produce healthy communities. For the quarter century or so after World War II a superficial analysis might even have found support for such a proposition. American forest historian Paul Hirt argues that during this era a "conspiracy of optimism" governed National Forest management in the western United States. British Columbia fell under the same spell, marked by a faith in private enterprise to provide technological answers to social and environmental problems. The great flaw of sustained-yield forestry as embraced in this province is our assumption that the government need only allocate timber, provide a positive climate for investment and then withdraw to a distant supervisory role.[1]

But British Columbia's problems run deeper than that, for even a more rigorous regulatory regime in the area of forest renewal would not have addressed the fundamental inequities of forest law and practice in this province. The most basic and tragic of these rests on the separation of First Nations from their land without benefit of treaty. A society based upon Euro-American settlement and capitalist enterprise dictated that from the outset land and resources would be allocated to those deemed able to extract the highest value from their market potential. That comparatively little violence occurred in this process should not obscure the ruthless way in which aboriginal people were deemed obstacles to progress and pushed aside. We are only now beginning to confront this reality in a way that offers even a thin commitment to fairness.

Consideration of the place up-coast forests held in the economic life of the province presents a second dimension of inequality. Policy makers from the turn of the 20th century devoted far too little attention to ensuring that these forests sustained the region's communities. The fact that the protected waters between Vancouver Island and the mainland offered such a convenient route to funnel northern resources south simply reinforced the logic of this traffic. But neither was it inevitable that metropolitan centres would capture so much of the region's potential wealth. District foresters periodically raised the issue with Victoria-based superiors, but to no avail. The Colcel development at Prince Rupert provided an opportunity for regional processing of northwestern timber, and for a couple of decades Tree Farm Licence No. 1 seemed to serve its purpose. More research is needed to disentangle the problems that eventually brought that enterprise down, but there is no denying that its enormous scale never equated to a solid foundation for communities in the region. After World War II the creation of Tree Farm Licences consisting of discrete blocks spread out over the entire coast made a mockery of the supposed link between sustained-yield and up-coast community development.

The pulp-and-paper towns of Ocean Falls and Prince Rupert present clear examples of yet another source of inequality. The enormous power wielded by these operations created regional hinterlands, all subject to exploitation at the hands of the corporate entity. Pacific Mills abused its log suppliers because its reach extended over a wide territory without the constraints a more diversified industrial structure would have encouraged. Columbia Cellulose and its successors threw a similar sort of monopolistic net over the Skeena and Nass watersheds. In the end, neither enterprise achieved the sort of permanence that might have made the trade-off in economic and social equity remotely worthwhile.

And so, we are left to contemplate the different paths that might have been taken. So much of our industrial history makes depressing reading, involving concentrations of power that seemingly reduce people to unwitting cogs or abject victims. But up-coast forest history exhibits vivid stories of resistance as well as accommodation. First Nations suffered, but not silently. They became involved in the forest industry for their own reasons even as they challenged its control over the landscape. Woodworkers protested their working and living conditions, succeeding in building the province's largest union to counter the power of capital. Residents of Bella Coola, Hazelton and other communities never accepted meekly the influence exerted by the multinationals. It is perhaps from these threads of dissent that a new, brighter fabric of up-coast forestry can be woven.

1 Shallow Roots

Early Forest Industrialization, 1880–1914

Forest exploitation for domestic and international markets occupied a relatively minor place in the up-coast economy prior to World War I. Fur and then salmon ranked above wood in the capitalist penetration of aboriginal space, and missionaries sought a harvest of souls at several points. Coastal First Nations, whose connection to nature embodied a commitment to place and an ethic of reciprocity inspired by spirituality, fell under colonialist structures of domination after British Columbia's entry into Confederation. Reserves established along the north coast for the Tsimshian and Nisga'a during the 1880s ignored aboriginal systems of property rights and threatened access to traditional hunting and fishing sites. After a brief period of "relative generosity" under Indian Reserve Commissioner Gilbert M. Sproat in the previous decade, successor Peter O'Reilly proceeded to lay out reserves in a miserly spirit consistent with that of his father-in-law, Joseph Trutch.[1]

As for the more fundamental question of aboriginal land title, the new province retained its colonial heritage of outright opposition. Nor would north coast First Nations find an ally in Ottawa on this issue. Beset by acrimony in its relations with British Columbia, and concerned about the expense of negotiating treaties to extinguish aboriginal title, John A. Macdonald's Conservative government allowed the "spoilt child of Confederation" to have its way. Nisga'a and Tsimshian dissatisfaction with O'Reilly's hastily imposed reserves led to an 1887 Northwest Coast Commission to hear grievances, but both federal and provincial representatives were hostile to any notion of First Nations land ownership. By the 1880s, Robin Fisher points out, the reciprocal relationships of the fur trade had long since given way to the conflict of settlement and industrial interests determined to wrest control of land and resources from First Nations.[2]

At the same time, recent scholarship makes clear that First Nations played a crucial role in British Columbia's emerging industrial economy. This is no less true of forestry than the more well-known example of the salmon fishery. From the 1880s to World War I the forest industry, already well established on the Lower Mainland and Vancouver Island, made its initial inroads north of Seymour Inlet. It did so in a diverse fashion, supplying building materials and meeting the needs of salmon canneries. Not until after the turn of the century did the provincial state's promotional aspirations produce tenure policies that made the spruce, hemlock and cedar forests attractive investments and draw large-scale enterprise to Swanson Bay and Ocean Falls. Pulp-and-paper manufacturing rather than sawmilling would spark up-coast industrialization, locking up timber in corporate holdings and failing to deliver on immediate promises of stable employment and community development.

Indeed, by World War I, outside interests had seized control over huge tracts of up-coast timber. Much of this activity was purely speculative in nature, a capital investment with no regard for the forest's capacity to fulfil the needs of aboriginal and non-aboriginal inhabitants much less the habitat requirements of other terrestrial and aquatic species. Business failures and an inability to penetrate commodity markets made for a lacklustre start on the part of the early pulp-and-paper industrialists. But they, at least, came with visions of development that matched the goals of revenue-hungry policy makers in Victoria. The railway, the great engine of industrial capitalism in the late 19th and early 20th centuries, offered yet another strategy to link northern resources to consumers. As British Columbia prepared to send its sons to war, construction of the Grand Trunk Pacific Railway to the new town and port of Prince Rupert seemed certain to create a prosperous future for the northwest.[3]

Captain James Cook initiated non-aboriginal use of the forest resource during his third Pacific expedition in 1778 when his crew cut new Douglas-fir spars from stands at Nootka Sound. Furs, not timber, attracted American and European traders to the coast after the publication of Cook's account in 1784. From 1792 to about 1820 the maritime fur trade drew coastal First Nations into the world economy, although merchants such as John Meares carried spars to China along with more valuable cargoes of Sea Otter pelts. The establishment of North West Company and Hudson's Bay Company forts created a small demand for lumber, and six years after the two companies merged in 1821 a water-powered mill went into operation at Fort Vancouver. By the time the new Hudson's Bay Company moved to Fort Victoria in 1843 lumber exports to Hawaii had already begun, supplementing the trade in furs. During the 1850s, mills at Esquimalt and Nanaimo cut for local demand, with the former facility also producing lumber for export to Hawaii and California.[4]

Still, it was the large Puget Sound mills that provided much of the lumber for Victoria's brief construction boom during the Fraser River gold rush. After Edward Stamp's short-lived effort to initiate mass-production lumbering at Alberni to meet expanding Pacific Rim markets in the early 1860s, Burrard Inlet took centre stage. Stamp himself relocated to the south shore of the inlet, building a mill with English backing in 1865 that San Francisco interests took over two years later, establishing the Hastings Sawmill Company. On the other side of the inlet, Sewell Moody's Moodyville operation went into production in 1868. Mills at Chemainus and New Westminster supplemented this emerging industrial picture, but over the next couple of decades Burrard Inlet remained the sole important coastal lumbering centre.[5]

These mills assumed control over timber within a colonial context of inherited British traditions that emphasized Crown retention of forestland. Cutting rights would be granted through a variety of leasing and licensing schemes, with the Crown collecting resource rents in the form of ground rentals and royalty charges on the timber harvested by private enterprise. When a crushing debt load generated by infrastructure expenditures during the gold-rush period forced British Columbia to join Confederation in 1871, the new province retained ownership of its land and resources. By 1890, four years after the commencement of regular Canadian Pacific Railway service to Vancouver integrated the province into the Canadian economy, 41 sawmills operated in the coastal region, most drawing timber from long-term leases that imposed fixed annual rentals and royalties, the latter in theory subject to change at the government's discretion. Independent loggers had access to special timber licences, tenures that offered smaller tracts. Regulations from 1888, designed to prevent speculation in Crown timberlands, limited each holder to a single renewable but non-transferable licence.[6]

The adoption of the handlogger's licence in 1888 brought these colourful figures into the Crown's revenue stream. The proliferation of long inlets, bays and islands fronted by steep hillsides along British Columbia's coastline permitted the evolution of this sector on Burrard Inlet as early as the 1860s, consisting of hardy loggers who, with the aid of axes, saws and jacks, dropped timber directly into protected waters for booming. They gradually worked their way up the coast in search of suitable stands, paying a $10.00 annual fee in addition to royalties.[7]

By 1890, then, the provincial government had erected a basic system of tenures intended to encourage industrial expansion and capture rents from Crown forests. But these arrangements had little significance for the central and north coasts, although Lower Mainland operators had already begun looking northward for accessible, high-quality timber. John Hendry led the way after integrating both the Hastings and Moodyville operations into his new B.C. Mills Timber and Trading Company in the late 1880s. Since steep ground on Howe Sound and the Sechelt peninsula prevented efficient logging with oxen and horse teams, East Thurlow Island, Quadra Island and Rock Bay, just north of Campbell River, became the first "up-coast" sites of mass-production logging. Tugboats moved the logs from these sites to Lower Mainland mills, a transportation system that also serviced the growing handlogging sector.[8]

Claxton Sawmill, 1892.
BC Archives F-06160

During the following decade, larger operators initiated the industrial revolution in coast logging, adapting the steam-hoisting engine to yarding operations. And by the turn of the century, larger and more powerful steam "donkeys" were increasingly linked to industrial railroads at the largest camps. Mechanized logging reached new heights of sophistication over the next decade with the introduction of overhead yarding featuring massive steam engines that brought logs to railroad landings over complex cable systems rigged to spar trees. Faster, more hazardous for logging crews, and suitable only for clearcutting, such technologies left "a trail of desolation" in their wake.[9]

During the late 19th century, industrial logging and sawmilling had only a marginal impact on north-coast aboriginal populations and ecosystems. The Hudson's Bay Company's northern posts provided small pockets of non-aboriginal settlement, and the Cassiar gold rush attracted a couple of hundred miners. According to R. Cole Harris and Robert Galois, 2,893 Tsimshian and Nisga'a occupied the lower Skeena and Nass rivers in 1881. Just over 820 Haida lived on the Queen Charlotte Islands. Evangelical activities at the Metlakatla, Port Simpson, Greenville and Kincolith missions, and commerce at the Hudson Bay Company's fur trade post at Port Simpson accounted for the presence of most non-aboriginal people on the north coast. Of growing economic and social importance were the salmon canneries that relied largely on aboriginal and Chinese labour. The first opened on the Skeena in 1877, then on the Nass in 1881. By 1890 the two rivers hosted 11 canneries. Another 6 had been established to the south at Rivers Inlet, and 13 others operated elsewhere along the central coast.[10]

Coastal aboriginal people had begun travelling by canoe to work at the Fraser River canneries with the establishment of the first plants there in the

Claxton Cannery, 1890s.
BC Archives F-06161

early 1870s, fitting wage labour into their seasonal resource-use patterns. The appearance of up-coast facilities quickly drew local First Peoples to these sites in the summer months. The Tsimshian people along the Skeena began gravitating to the canneries, according to James McDonald. Hartley Bay families spent the summers at Rivers Inlet, Lowe Inlet or the Skeena River, with men fishing and women working on the canning lines. Rivers Inlet cannery operators dispatched tugboats to the Haisla village of Kitimat each June, towing long lines of canoes back to the canneries to fill their seasonal labour demands.[11]

Coastal First Peoples were no strangers to the time and work discipline of industrial capitalism by the late 19th century, then, often devoting their earnings to the staging of potlatches and other traditional cultural activities. Employers who required a stable, year-round labour force expressed frustration at the ceremonial winter gatherings that drew aboriginal workers away from their plants for, contributing to adoption of the 1884 potlatch ban. Enforced with mixed success, the law nevertheless reflects the importance government, economic and religious figures attached to the cultivation of what John Lutz terms the "stable habits of industry" considered essential to the "development of a Christian capitalist society".[12]

The history of forest exploitation along the north coast prior to the 20th century is murky, although it seems clear that early missionary activities, salmon canneries and the non-aboriginal settlement process figured prominently in the appearance of the first sawmills. William Duncan's model village at Metlakatla, established in 1862, hosted a range of industrial pursuits by the late 1870s. A sawmill, shingle mill and sash-and-door factory were among the enterprises that failed to produce returns during their limited period of operation. When conflicts between Duncan and the

Church Missionary Society led to his departure for Alaska with a group of followers, some of Metlakatla's inhabitants moved to Hartley Bay. To support a mission at that point, Methodists sponsored the construction of a small water-powered mill in 1889 to provide materials for church and village construction. By around 1900, aboriginal bands and missions operated mills at a number of north coast points, including Masset, Skidegate, Kincolith and Andimaul.[13]

The first commercial sawmill on the north coast was likely the Georgetown Mill, near Port Simpson. In operation by 1874, the plant counted Tsimshian workers among its loggers, mill crew and longshoremen. The Cunningham Mill at Port Essington also began cutting prior to 1895, and it is likely that both plants filled orders from Skeena River canneries for wood that went into the construction of packing boxes. First Peoples along the upper Skeena cut cordwood for the salmon canneries in addition to seasonal fishing and cannery work, hunting and trapping. Down the coast at Bella Coola, the arrival of a group of Norwegian settlers from Minnesota led to the establishment of a sawmill by 1898, and it is probable that cannery development at Rivers Inlet was accompanied by at least one small mill to cut box material. By the end of the 1890s, according to Knight, logging in the Department of Indian Affairs' North West Coast Agency, extending from Rivers Inlet to the Nass, "ranked second only to commercial fishing and cannery work as a source of Indian cash income." Summarizing aboriginal economic life in that agency in 1902, Indian Agent C. Todd described the principal occupations as:

> salmon-fishing and canning, procuring saw-logs for the three sawmills of the northwest coast, the eulachon fishing industry, hunting and trapping, fur-seal hunting, procuring and drying herring spawn, catching and drying salmon and halibut for food, boat and canoe making, and cutting fire-wood for the use of the salmon canneries.[14]

Settlement, canneries and missionary work had combined to give logging and sawmilling a minor place in the north coast economy by the turn of the century, but the industry had no connection to export markets. The small mills would have operated intermittently, most likely securing logs from the sporadic efforts of the handloggers who supplied the Cunningham and Georgetown plants as demand for local construction and packing material warranted. Indeed, in 1901 the provincial forest industry contributed just seven per cent of government revenues. But change was in the offing as the recession of the early 1890s gave way to an early 20th-century economic boom, and governments showed a new commitment to promote forest industrialization.[15]

A series of amendments to the Lands Act in 1901 signalled the provincial government's determination to attract investment capital, stimulate forest sector employment, and bolster the flow of revenues to the treasury. Timber leases were made renewable for successive 21-year terms, and while the renewable one-year special timber licences remained non-transferable, individuals or companies were permitted to hold two of the tenures. Two years later, government responded to industry's demands for greater tenure security by extending the duration of these licences to five years. The imposition of the 1901 "manufacturing condition" on all timber cut from leased and licensed lands sought to limit the export of logs to Puget Sound mills. These measures would all have implications for the north-coast forest industry, but pale in significance to the government's effort to encourage investment in pulp-and-paper manufacturing. With vast expanses of spruce and hemlock stands awaiting exploitation, but not a single pulp mill to realize their value, the government amended the Lands Act to make available renewable 21-year pulp leases at an annual rental of two cents an acre, low royalty rates and water-power rights to corporations that committed to build and operate a plant by the end of 1909.[16]

The terms were attractive, to say the least. Annual timber-lease rentals amounted to 22 cents an acre at this time, and special timber licences charged 15 cents. Holders paid a royalty of 50 cents per thousand feet on both, in contrast to the 35-cent levy offered prospective pulp-and-paper producers. But such plants required much larger capital outlays than sawmills, demanding more generous inducements. Four firms responded, quickly seizing control of over 350,000 acres of timberland. The Canadian Industrial Company acquired the rights to 134,551 acres at Powell River. The Quatsino Power and Pulp Company secured 55,669 acres on northern Vancouver Island. Groups of British and American capitalists claimed vast stretches of Crown timber further north on the mainland coast. The Oriental Power and Paper Company seized 84,180 acres at Swanson Bay. Between Swanson Bay and Powell River, Seattle interests organized the Bella Coola Pulp and Paper Company to stake claim to some 80,000 acres on South Bentinck Arm, Dean Channel, Burke Channel and Cousins Inlet, deciding to locate a mill at Ocean Falls. By 1904 the government and promoters of the four ventures agreed that "sufficient inducements had been offered for the pioneering of the new industry," leading to repeal of the legislation.[17]

Although all of the above companies proved unable to realize their goals, the Ocean Falls and Swanson Bay leases did ultimately provide the foundation for pulp-and-paper projects, albeit of quite different durations. Both would be plagued by early problems that produced a flurry of ownership changes, aborted developments and dismal financial performance. But in the interim, Richard McBride's Conservative government responded to a 1904 slump in the lumber market by meeting the timber industry's demands for more changes to the tenure system. The amendments would have far-reaching consequences, opening the door to a wave of speculative capital investment that saw much of the province's best and most accessible forestland come under corporate control in a brief, two-year frenzy of timber staking.

A. Stavdahl and M. Hammer lived and worked in the Bella Coola River valley
in the 1890s.
BC Archives G-00977

Unable to secure bank loans on their special timber licences in 1904, logging operators renewed their demands for greater tenure security. Anxious to bolster investor confidence and raise forest revenues, McBride moved early the following year to adopt sweeping amendments to the Lands Act. Existing licences would now be renewable for 16 years, and those acquired after April 1905 for 21 years. Both categories of licences were made transferable, with no limit on individual holdings. With the stroke of a pen McBride had eliminated many of industry's concerns about the security of their tenures, thrown down the welcome mat to speculators and declared North America's last accessible timber frontier open to outside capital on appealing terms. Perfectly timed to take advantage of timber depletion in eastern areas of the continent and mounting demand for building materials, McBride's revised licences attracted unprecedented interest. Between 1905 and the late 1907 moratorium, cruisers staked out over 15,000 licences, covering roughly 12,000 square miles.[18]

By this time, handlogging had assumed growing importance for First Peoples along the north coast, complementing other seasonal pursuits. At Kitimat, Hartley Bay, Bella Bella, Bella Coola and Kimsquit some aborigi-

Mark Smaby's first cabin, Ocean Falls, 1908.
BC Archives I-50603

nal men devoted the fall and early winter to logging for local mills, but as the timber-staking boom extended into up-coast forests opportunities to accumulate income from this source declined. In 1907, they complained to Indian Agent George Morrow: "Owing to the great rush for timber limits during the year, it is not an easy matter for the handlogger to make this class of work profitable, as the best of the timber has all been taken up."[19]

Revenue from rental fees helped meet McBride's immediate financial needs, but the up-coast pulp-and-paper projects made only halting progress. At Ocean Falls the original enterprise was reorganized as the Bella Coola Development Company, and after two years of timber cruising and surveying the promoters gained government approval of their pulp lease in 1906. Land clearing at the head of Cousins Inlet began that autumn with a small crew that travelled to the prospective mill site by tugboat from Bella Coola, the nearest stop on the CPR's coastal steamship service. Lack of capital stalled this work, and in 1907 only Mark Smaby and two other men continued clearing the site that would become the community of Ocean Falls. Settlers at Bella Coola expressed frustration at both the tying up of so much land in pulp leases and the lack of development. But the prospects for progress dimmed further with the filing of a lawsuit by investors against the two Seattle principals. Unable to meet their obligations, they sold out to a group headed by Seattle lumberman Lester David, who organized the new Ocean Falls Company in 1909. Directors included I. Hamilton Benn and J.S.F. Lawson of England, Quebec paper magnate William Price,

Ocean Falls construction crew, 1910.
BC Archives I-50604

Henry J. Crocker of San Francisco, and Vancouver banker J.H. Campbell. The *Western Lumberman* described the syndicate as "the strongest combination of capitalists yet interested in any industry in the province."[20]

The optimism seemed justified when David secured a $2 million bond issue to finance construction, and by 1909 a small sawmill had begun cutting lumber for wharf, town site and power-dam construction. Handloggers provided some of the logs, and in 1910 Smaby established the company's first camp on South Bentinck Arm with 30 men and two steam donkeys. By the summer of 1911 an estimated 500 workers were busy at Ocean Falls, and freighters docked frequently with equipment for a larger sawmill and the pulp mill. A hotel provided lodging for construction crews, a few houses had also been built, and a small hospital dispensed medical care. A large company store, a ubiquitous feature of British Columbia resource towns, sold tobacco and other essentials.[21]

At Swanson Bay, roughly equidistant between Prince Rupert and Ocean Falls, and about 15 miles from the fishing village of Butedale, the Oriental Power and Paper Company had spent an estimated $40,000 in acquiring and cruising its timber limits on the mainland and Princess Royal Island by the end of 1903. Surveying continued in 1904, with the firm's British backers predicting the erection of a 100-ton per day capacity paper mill within two years. But by early 1905 the company, a subsidiary of the Canadian Finance Syndicate of London, had apparently given way to the Canadian Pacific Pulp and Paper Company. Oriental Managing Director J.M. MacKinnon retained his post, returning from a three-month trip to London in that January to announce that the company would delay the building of a paper mill, but proceed immediately with a sawmill and 35-

40-ton-per-day sulphite pulp plant. Almost another year would pass before MacKinnon reported that the funding was finally in place to commence development. By mid-1906 the Oriental leases and capital stock belonged to the Canadian Pacific Sulphite Pulp Company, purchased for $187,000.[22]

The firm's London directors announced their intention to commence immediate development of the sawmill and pulp plant, predicting start-up sometime in 1907. By that June, the sawmill was under construction, and MacKinnon now envisioned production by the fall of 1908. Boilers and structural steel had not yet arrived at Vancouver in February, but the pace picked up after that, and on June 29, five carloads of machinery arrived by freighter. By August the tall digester building, acid towers, boiler house and other structures were nearly ready. The plant, wharf and workers' houses, some 400 miles north of Vancouver, would "combine to make a settlement that is a surprise to people taking the trip north," observed a writer. London director Ralph Reed, one of England's largest paper manufacturers, returned to Vancouver from a visit to Swanson Bay in late 1908 full of optimism. "I think that there is no property that could have more advantages," he declared, citing accessible timber, ample water power from a nearby waterfall, and easy access to shipping from the tidewater mill.[23]

The arrival of five American-made digesters in 1909 finally put the Swanson Bay pulp enterprise on a production basis. The sawmill was already running two shifts, filling orders for Prince Rupert contractors that summer in addition to cutting clear spruce for a Glasgow shipbuilder. Finally, in late summer or early fall, the Canadian Pacific sulphite plant went into operation after 18 months of development. The 25-ton-daily-capacity plant, described as "one of the finest pulp mills on the American continent," operated "night and day" by the end of the year. One of the first shipments of pulp went to the new British Canadian Wood Pulp and Paper Company paper mill at Port Mellon. Another larger cargo crossed the Pacific to Kobe, Japan. Most of the cedar and spruce logs cut at the Swanson Bay sawmill between 1906 and 1909 came from handloggers, going into the manufacture of lumber and box shooks sent on scows to Prince Rupert, Vancouver and Stewart. Troubles lay ahead for the Swanson Bay mill, but its opening must have pleased the McBride government, which issued notices to the other pulp lease holders that retention of their timber rights depended upon the commencement of production by November 30, 1911.[24]

The Swanson Bay operation itself soon floundered, closing down early in 1910 after producing about 1,000 tons of pulp under "conditions of rather intermittent operation". Another corporate reorganization followed, and by June 1910 the new Swanson Bay Forests, Wood Pulp and Lumber Company had acquired the Canadian Pacific assets. The new shareholders authorized expenditures to increase the capacity of the sawmill, pulp plant, and logging operation, but made no mention of going into the manufacture of paper. Late that year, in a statement meant to inspire confidence in the

Mills at Swanson Bay, 1910.
BC Archives F-06347

new enterprise, director J.W. Robson explained that output of the pulp mill would be increased from 5,000 to 7,000 tons per year, and daily lumber production to 80,000 feet. "The quality of our spruce limits enables us to make a pulp which is equal, if not superior, to the best production of Eastern Canada and Scandinavia," and the tidewater timber limits ensured "easy and economical logging operations".[25]

After a lengthy shutdown to complete alterations, the sawmill and pulp plant resumed operation in early September 1911, with about 200 workers. The new band mill broke down the large spruce logs into sizes suitable for the pulp mill, and processed the clear spruce into lumber. The company reported several orders for cooked pulp from the American market, although the mill failed to operate at capacity that autumn. At least one shipment of clear spruce lumber went to Vancouver for transfer to a Blue Funnel liner bound for Liverpool, but reports at the end of the year indicated that the firm would default on its first mortgage bonds. The operation passed into the control of the Vancouver firm of Evans, Coleman and Evans shortly thereafter, and apparently shut down. The plant sat idle in early 1913, and there is no record of operation over the next two years. In 1915, with the plant "believed to be in very bad repair", Vancouver and London parties launched legal action to secure an accounting of the Swanson Bay Forests, Wood Pulp and Lumber Mills administration.[26]

While Swanson Bay slipped back into isolation, populated only by a few handloggers who produced booms for Prince Rupert mills and box making at the Namu cannery, prospects for sustained operation at Ocean Falls had also dimmed. Lester David's Ocean Falls Company had managed to commence the production of mechanical pulp and lumber by the end of

Ocean Falls town and mills, 1912.
BC Archives I-50579

1912. Over 30 houses had been built, the concrete dam had 10,000-horse-power capacity, but development outpaced David's ability to meet creditors' demands. Facing debts totalling $300,000, David resigned as president at the end of 1912. English shareholders gained control of the company, then shut the operation down the following May, throwing 400 employees out of work. A three-year "period of depression" ensued for north-coast handloggers, who now had neither the Swanson Bay nor Ocean Falls mills as markets.[27]

While the early north-coast pulp initiatives sputtered along under mounting debt and shifting ownership, construction of the Grand Trunk Pacific Railway (GTP) transformed settlement patterns and economic life in the Skeena River region. Seasonal fishing and canning, the fur trade, sporadic mining activity, and scattered homesteading provided little impetus for permanent non-aboriginal settlement when the federal and provincial governments of Prime Minister Wilfred Laurier and Richard McBride undertook to open the north's resources to large-scale exploitation by supporting the construction of two new transcontinental rail lines. Grand Trunk Railway general manager Charles Hays hatched the plan to extend the firm's main line from North Bay across the prairie wheat belt to Port Simpson in 1902, while William Mackenzie and Donald Mann were piecing together their Canadian Northern Railway network. Failing to bring the two firms together in a collaborative project, Laurier allowed them both to go ahead. Incorporated on October 24, 1903, the GTP would press west from Winnipeg and east from the coast. After preliminary surveys, the company announced that Kaien Island, about 25 miles south of Port Simpson, would be the western terminus.[28]

Ties for the Grand Trunk Pacific Railway, 1913.
BC Archives NA-03949

Cruisers swarmed into the valleys of the Upper Fraser, Endako, Nechako, Buckley and Skeena rivers in the wake of the GTP's announcement and McBride's new tenure policies. Land clearing at the GTP's settlement of Prince Rupert began in 1906, after McBride facilitated its purchase of land at that point. The first rails were not laid for two years, but the GTP began contracting for the delivery of ties as early as 1907. The Georgetown mill, the B.C. Tie and Timber plant at Seal Harbour, and the Cunningham mill at Port Essington had difficulty furnishing the lumber required for development at Prince Rupert and the building of construction camps, leading outside capital to get in on the flourishing tie and lumber trade. In 1908, Seattle and Vancouver interests undertook construction of a mill near Aberdeen after acquiring 20 million feet of timber in the area, and a group of Seattle and Vancouver investors opened a plant at Skeena City that could produce 50,000 feet per day. Despite the orders for more lumber than local mills could turn out, the Seal Harbour mill and steam-logging operations at Alice Arm closed that autumn, shortly after the Cunningham plant at Port Essington burned down. Unable to pay its Japanese workers, the B.C. Tie and Timber assets were put up for sale after the sheriff seized the property, but the government stepped in to stop the sale until its claims for royalties on stumpage from the company's Alice Arm limits were satisfied. That mill too went up in flames on December 4, but the Prince Rupert Sash and Door factory opened for business at about the same time. Fires sparked by contractors along the GTP right-of-way also began taking their toll on the Skeena River's forests that fall.[29]

Small mills appeared along the Skeena to meet GTP demand during this period. C.A. Lillesburg opened a plant at Kitsumkalum, and the Hardscrabble Lumber Company mill near Kitselas began cutting by the end of 1909. Timber rights also began to move in response to the promise of

George Little Sawmill, Terrace, 1910.
BC Archives E-01735

access to North American markets. A.F. Sutherland of Vancouver sold 93 limits on the Nass and Kitsumkalum rivers to a group of British and Canadian investors who announced their intention to build a mill in the Prince Rupert area and take part in both local and export trade. Seattle interests acquired 300 million feet of timber on the Copper River, and the first of many pulp-and-paper rumours began circulating at Prince Rupert when the Pacific Pulp and Power Company was organized in Montreal to operate pulpwood holdings on the Nass and Skeena rivers. An English syndicate purchased roughly 20,000 acres of Crown timber rights on the Skeena near Prince Rupert from the National Timber Company. "The opening up of the Skeena River country by the GTP is opening up a new timber region in British Columbia," observed the *Western Lumberman.*[30]

By 1912, Prince Rupert's population had grown to 6,500, and while the speculative boom in local real estate would soon collapse, town site development provided work for 25 employees at the Prince Rupert Sash and Door plant. But the company imported most of its lumber from Vancouver. The Skeena Lumber Company at Skeena City, formerly oriented mainly to filling cannery orders, began directing its seasonal cut to the building boom. "Settlers are now pouring into the district and the output of the mill will be required for construction of homes for newcomers," reported a trade journal. But orders failed to materialize due to the GTP's slow pace of construction, leading to the mill's closure until 1914 when John Clarke acquired an interest in the property.[31]

Farther up the Skeena, George Little provided land for a GTP station at what would become Terrace. Surveyed in 1910, the town site attracted a

Winter logging, Terrace, 1915.
BC Archives E-01788

number of homesteaders from the Kitsumkalum River valley. Little built a
mill there the following year, and to the west at Prince Rupert the boom
continued through 1912. "One is struck in Prince Rupert and elsewhere up
the line with the spirit of optimism which prevails," wrote one government
official. "Everyone is a firm believer in the future of the northern country."
Lester David won the contract to supply lumber for the drydock from
Ocean Falls, and planned to operate large car ferries to Prince Rupert,
bringing cars loaded with lumber for shipment up the GTP line. Rail traffic opened as far east as Smithers by the end of 1913, but the GTP was
already struggling under financial difficulties despite its provincial
subsidies.[32]

Across Hecate Straight, the extensive stands of high-quality Sitka
Spruce began attracting American development capital in 1908 with the
arrival of the Moresby Island Lumber Company. The mid-western
investors purchased options on 49 square miles of Crown timber and 8,000
acres of private timberland on Graham Island, another 40 square miles on
Moresby Island that spring, and announced that mill construction at the
new town site of Queen Charlotte City would commence immediately if
the GTP began pushing east from Prince Rupert in the summer. The firm
purchased logging donkeys and mill equipment from Vancouver suppliers.
Los Angeles interests behind the Graham Island Lumber Company, holding over 100,000 acres on northern Graham Island, came forward with
ambitious plans for Masset in 1908. Vice-President C.M. Shannon promised that a new mill would "be among the greatest on the coast", engaging
in both the domestic and export trade.[33]

The Moresby Island Lumber Company's machinery began arriving at

Sawmill at Queen Charlotte City between1910 and 1920.
BC Archives E-00089

Queen Charlotte City by steamship that fall. By the summer of 1910, two camps supplied logs for the plant, which had a capacity of 80,000 feet per day, cutting clear spruce and hemlock for the Australian market. Although the firm's Japanese boarding house at Queen Charlotte City burned down, the community's 400 residents had access to a hospital, school, hotel and drug store. Orders for lumber needed in the construction of an up-coast whaling station had the mill running overtime in October, and then came the news that the timber holdings and mill had been sold to another American syndicate headed by A.C. Frost of Chicago. The new owners apparently continued operating the mill under the same name, receiving an order in 1911 from the British Admiralty for 200,000 feet of spruce to be used in the manufacture of oars.[34]

Timber and coal properties on the Queen Charlottes changed hands with some frequency in the years before World War I, but few actual developments accompanied the transactions. The Graham Island Company finally shipped a small mill to Masset to cut lumber for a coal mining operation late in 1911, and opened a lumberyard there for purchases by settlers. That spring the new British Canadian Lumber Corporation, capitalized at $20 million and holding about 14 million feet of timber in British Columbia, announced plans to build a sawmill on Masset Inlet to cut its extensive Graham Island tracts. One optimistic report even hinted at the erection of a pulp mill on Masset Inlet, and later in 1911, General Manager M.F. Buckley said a large plant would be built at Prince Rupert. By mid 1912 only the Queen Charlotte City sawmill operated on the islands, but reports continued to circulate that British Canadian would erect its promised Masset Inlet mill shortly.[35]

Westholme Lumber Company, Prince Rupert, 1910.
BC Archives NA-41628

Elsewhere along the north coast, lumbering remained an uncertain undertaking, linked to the cannery trade and the halting process of local settlement. At Hartley Bay the original water-powered sawmill had fallen idle because of an inadequate water supply. Captain Edward McCroskie purchased land there in 1905 and established the Hartley Bay Lumber Trading and Fishing Company. McCroskie then built a 20,000-foot-capacity mill that operated on a seasonal basis with local aboriginal labour, cutting mostly Yellow-cedar until 1908 when Michigan capitalists acquired a controlling interest in the enterprise. McCroskie remained involved, but reports the following year indicate that A. Anderson of Victoria had assumed ownership. McCroskie's relationship to the operation from that point on is unknown. He was appointed the government's timber inspector for the Queen Charlotte and Skeena districts in 1912, and it seems certain that the mill had ceased operation by that time. When construction of a new wharf at Hartley Bay began in 1913, the Georgetown sawmill supplied the lumber.[36]

Bella Coola, originally mentioned as the site of the pulp mill eventually built 60 miles west at Ocean Falls, gained only a small water-powered sawmill around 1913 when five partners organized the Salloompt Mills at Hagensborg. Oxen or horses hauled the logs for the plant, a seasonal operation that logged only when snowfall facilitated log transportation. When the mild winter of 1912–13 allowed hauling over snow roads for only a brief time, local lumber supplies suffered. But the next winter provided better conditions. Residents welcomed a heavy snowfall in late January 1914, which saw logging operations go into "full swing" at Hagensborg. "The local mills have all available teams engaged, and there is little likelihood of

Sawmill, Georgetown, 1914.
BC Archives NA-04353

a shortage of lumber for local purposes next summer," noted the *Bella Coola Courier.*[37]

With both the Ocean Falls and Swanson Bay pulp plants shut down in the summer of 1913, logging on the north coast was entirely the domain of the handlogger. "I know of no logging operations on licences or leases in this district," Prince Rupert district forester H.S. Irwin reported. "All logging is now done by handloggers.' Their logs might go to one of 14 mills, none of which seemed to have operated continuously during this period. In 1913 Skeena River region mills included the Georgetown Saw Mill Company at Big Bay, the James Brown mill at Port Essington and the idle Skeena mill at Skeena City. At Prince Rupert the Westholme Lumber Company had a lumber yard, and an operation known as the Prince Rupert Planing Mills had emerged. W.L. Barton and W.E. Wanless conducted small sawmilling enterprises on Masset Inlet. The Findlay, Durham and Brodie mill at Rivers Inlet cut material for canneries there, though it went out of commission temporarily that fall when a fire destroyed the boiler house. Draney Fisheries operated a similar concern at Namu, turning out box lumber. At Hagensborg the Salloompt Mill and another owned by H.O. Hansen filled local construction requirements, and late that year, G.W. Smith of Ocean Falls reportedly took over a defunct Bella Bella plant with the intention of initiating production.[38]

Isolated from the Lower Mainland and Vancouver Island construction markets and lacking shipping connections to export destinations, the region's sawmillers placed their faith in rapid completion of the GTP to link their operations to the Canadian prairies and American Midwest. Alternatively, freighters might pick up cargoes at Prince Rupert for shipment to

Martin Grainger with forester J. Latham in Hazelton, 1914.
BC Archives H-05547

Asia. Perhaps, too, the soon-to-be completed Panama Canal would generate a water-borne trade to the east coast. Prospects were also brighter on
the pulp-and-paper front, at least at Ocean Falls. Pressure from newspaper
publishers was responsible for the elimination of American tariffs on
newsprint in 1913, and the giant Crown Willamette Paper Company enter-
tained overtures from Ocean Falls shareholders about taking control of the
idle operation.[39]

Timber interests around the province might also draw some solace
from the new Forest Act. In 1909 McBride had promised to meet industry's
demand for greater tenure security, pending the appointment of a Royal
Commission to make comprehensive forest policy recommendations. Riding the wave of the early 20th century North American conservation movement, Minister of Lands F.J. Fulton headed an inquiry that held hearings
around the province and interviewed American and Canadian conservation leaders, reporting to the Legislature in January 1911. McBride had
already followed through on the Commission's recommendation that existing leases and licences be made renewable in perpetuity, a decision that
placed his and future administrations in a cooperative relationship with the
major tenure holders. Fulton's final report expressed the same faith in busi-
ness-state cooperation to achieve rational forest management, proposing

the creation of a new agency to regulate logging, collect revenues and administer a forest protection scheme funded equally by industry and government.[40]

The McBride government moved quickly to introduce what new Minister of Lands William Ross called "a sane and businesslike policy of conservation free from sentimental extravagance." The 1912 Forest Act established a Forest Branch in the Department of Lands under Chief Forester H.R. MacMillan. MacMillan's agency would administer the shared-cost Forest Protection Fund, and a new timber-sale tenure designed to permit the logging of small tracts adjacent to the existing leases and licences. These old temporary tenures, where the bulk of the logging took place for several decades, remained outside the Crown's power to regulate logging practices, along with private lands such as Vancouver Island's Esquimalt and Nanaimo Railway grant. MacMillan organized the province into 11 forest districts, a plan that underwent subsequent modification. R.E. Allen took charge of the 13,786,000 acre Hazelton district, taking in the watersheds of the Skeena and Bulkley rivers. H.S. Irwin and a small staff administered the Prince Rupert forest district, an area of 18,723,000 acres west of the Cascade Mountains extending from the northern tip of Vancouver Island to Portland Canal.[41]

The government took its first, halting steps toward the scientific management of Crown timberlands just as Canada's early-20th-century economic boom began to wane. The largest Lower Mainland and Vancouver Island mills had neglected offshore trade in favour of the prairie market, which collapsed in 1913 due to crop failures and falling wheat prices. MacMillan termed 1913 a "year of financial stringency", despite record-breaking forest revenues of almost $3 million. He identified better fire protection and market extension as pressing needs, and pointed to the lack of adequate staff in coastal and northern districts, where rapid development and settlement demanded more attention to timber sale, land classification, timber inspection and protection work. On the bright side, newsprint production at the huge new Powell River plant, "a monument to commercial enterprise and Government policy", had created a "new outpost of white settlement on the vast stretch of the Provincial coast-line." But duty-free entry of pulp and paper into the American market had not yet revived the Swanson Bay or Ocean Falls mills. On the other hand, both the GTP line and the Panama Canal were nearing completion, promising new opportunities to link northern timber to east-coast markets. Or so it seemed, at least, as the world drifted toward a war that would have unforeseen consequences for the north-coast economy.[42]

An early handlogger on Powell River, 1910.
Rare Books & Special Collections, UBC Library, BC 1930/277/10

2 The Spruce Drive

World War I and the Forest Economy

War produces dramatic change as economies shift to meet the urgent demand for military materials. Social and environmental change follow as governments set new priorities, and nowhere would this pattern be more evident than on the Queen Charlotte Islands during World War I. Before 1916, the islands' magnificent stands of Sitka Spruce, isolated by the hazardous waters of Hecate Straight and a prohibitively long tow to Lower Mainland mills, had marginal commercial value. The war changed that almost overnight, joining the spruce's special qualities to the needs of Allied military aviation. The result was a brief but intense assault on the forests around Masset Inlet and other points, one that left an unquantifiable legacy of waste but little in the way of lasting industrial infrastructure. Sustainability, both economic and environmental, had little importance to the nations at war or to business interests seeking to capitalize on the opportunities of artificial wartime markets.

Looking past the frenetic pace of the spruce drive, regional promoters had additional grounds for optimism during the war. Commencement of Grand Trunk Pacific Railway (GTP) service and the opening of the Panama Canal seemed certain to open distant markets to northern resources. Integration of Ocean Falls into the vast American-owned Crown Willamette organization gave that operation a solid foundation. The outlook for Swanson Bay even brightened for a time, although that would prove illusory. The lure of continental and Asian markets drew J.S. Emerson into the sawmilling business with a modern plant at Prince Rupert, but that enterprise, too, would encounter problems after the war. One would have had to look even harder to detect the most enduring feature of the emerging forest economy, the ongoing marginalization of First Nations at the hands of a timber-rights system that denied them access to the resource.

The onset of the war in Europe in August 1914 produced calamitous results for a forest industry already weakened by inflation and American competition in slumping prairie markets. Disruption of international shipping helped bring the lumber industry to a "dead halt". Orders for pulp and paper picked up unexpectedly, generating brisk business for the Powell River and Howe Sound mills despite shipping problems. On the north coast, GTP construction and ongoing development at Prince Rupert helped the local mills avoid the worst consequences of the 1913–14 slump, and orders from the 42 canneries and five cold storage plants remained strong. Indeed, salmon canning would enjoy a wartime boom due in part to the British government's decision to make canned salmon a staple of military rations. Completion of the GTP in April 1914 led to the commencement of passenger service between Prince Rupert and Winnipeg in September, but the coast city's boom was already over. The war curtailed the arrival of immigrants to the region, enlistments led to the departure of many, and Prince Rupert's population began to drop. But the railway would continue to generate a strong demand for local forest products. The Prince Rupert Board of Trade expressed optimism in early 1915 that the Panama Canal would facilitate pulp shipments, but complained of inadequate port facilities that hindered the ability of northern lumber mills to compete with southern producers. The lumber market remained "purely local" at the end of 1914, then, driven by the canneries, the GTP and small communities in the region.[1]

Completion of the drydock and ship-repair yards by a GTP subsidiary in 1915 held out the prospect of vessel construction orders, and GTP rail rates appeared low enough to permit the region's lumbermen to compete with Vancouver mills in the Alberta market, Prince Rupert Commissioner of Publicity F.S. Wright reported in the spring of 1915. Participation in the lumber markets of the eastern seaboard and Great Britain through the Panama Canal had proven uneconomical due to the long distance and absence of dry kilns in local mills, but the region's cedar poles would be attractive to American telephone and telegraph companies. Local and North American demand, Wright predicted, would "enable the lumber industry of this district to obtain a satisfactory start on the road to permanency."[2]

But as local development slowed, and outside markets did not appear, the industry along the Skeena began to contract rather than expand. The Prince Rupert Sash and Door factory went into liquidation early in 1915, and John Clarke shut down the Skeena River Lumber Company plant later that year, selling it in 1916 to satisfy a mortgage held by J.S. Emerson. Slack demand and low prices convinced backers of a Prince Rupert sawmill project to abandon the plan until conditions improved. Among the bright spots in 1915 was the traditional demand for boxes from the salmon and halibut processors, although southern manufacturers held about 80 per cent of this trade. The GTP, now teetering on the edge of bankruptcy, also continued to need piling, posts and ties for shipment to points along the line. More than a hundred handloggers had licences to cut isolated patches of tidewater timber, but many were inactive.[3]

Because handlogging remained so vital to north coast lumbering,

Early handlogging camp on Powell River 1910.
Rare Books & Special Collections, UBC Library, BC 1930/278

administrative matters drew critical commentary from everyone concerned. The Fulton Commission had recommended abolition of the
licences, citing the wastefulness of these operations, the alleged tendency of
handloggers to cut timber from other tenures and unalienated Crown land,
and the difficulty of regulating these transient loggers. The 1912 Forest Act
had not incorporated this proposal, but by 1914, delays in the application
process had become intolerable. "The handlogger in this neighbourhood
has had extreme difficulty in obtaining the necessary licence," observed the
Bella Coola Courier. Many didn't know where to apply, and the government
agent at Prince Rupert did not handle applications. The appointment of an
official timber scaler for the area in 1914 did not stop complaints about the
ponderous process of defining limits and obtaining licences.[4]

Fully aware that a month often elapsed between the submission of an
application and its approval in Victoria, District Forester H.S. Irwin cir-
cumvented the bureaucracy and allowed handloggers to operate before
their licences had been issued. The sawmills financed most of the loggers,
including their $25 licence fee, he explained, and had been slow forward-
ing payment in 1914. But it would be an easy matter to collect the overdue
fees from the mills, and strict adherence to the rules would delay the
issuance of licences for months, depriving the mills of their log supply. Victoria took a dim view of Irwin's common-sense approach, however, and
instructed him to cease allowing any handloggers to operate until they had
received their licences and paid the fees. Aware that delays created hard-
ships, and no doubt wishing to quiet criticism of a timber allocation policy
that had allowed speculators to lock up much of the province's commer-
cially accessible forest land, the Forest Branch took steps to accelerate

licence processing. By the end of the year most were mailed within three days of receipt of payment, MacMillan declared.[5]

While no doubt welcome, procedural improvements in the handlogging sector did not address the more fundamental political question in a hinterland region that had seen its resources turned over to outside interests content to hold them for speculative gain. "The bulk of the land sold to speculators lies in the northern portion of the province," asserted the editor of the *Bella Coola Courier* in a rebuke of Richard McBride's administration. "Northern British Columbia has been exploited and has received no adequate return. We were once rich and had we been separate from the South, a province apart, we would be rich today. As it is, we have very little left to us of our once vast natural resources."[6]

But if non-aboriginal people along the north coast felt a sense of grievance at the sight of timber held for future exploitation by those with no clear stake in the region, their frustration could not approach that of First Nations. As early as 1887, leaders of north-coast First Nations had approached the provincial government about land claims, inadequate reserves and access to Crown timber and other resources. The Nisga'a, active since the 1880s in protesting the province's unwillingness to accept the concept of aboriginal title to the land, had begun turning settlers away from the Nass River valley in 1907. In response to GTP construction along the Skeena River a delegation of northern chiefs went to Ottawa, presenting a petition to Prime Minister Wilfred Laurier. Laurier's promise of action to protect their interests, and a 1910 visit to Prince Rupert as part of a western swing produced further meetings, but the province's refusal to consider treaty negotiations led only to establishment of the McKenna-McBride Royal Commission of 1913. Lacking a mandate to consider the issue of treaty rights, the Commission concentrated on reserve allocations around the province.[7]

Nevertheless, up-coast First Nations took the opportunity of the Commission hearings to voice their resentment of forest policies that had deprived them of access to the resource. "All up and down the salt water there are posts saying that this land belongs to the white men who have bought it from the Government," declared a Bella Coola man. "If I now take any sticks of timber from these places the white men will come along and say 'leave that alone, it belongs to me.'" The chief of the Kimsquit band asserted that aboriginal people had few opportunities to log. "Most of the land is owned by white men who have bought it for speculation and there is no logging done on it now." Kitimats complained of the $25 handlogging licence fee, an amount that exceeded their financial resources. The Hartley Bay band asked for Gill Island to be set aside as a timber reserve for their use.[8]

In the face of such complaints, timber holders took steps to protect their interests against aboriginal encroachment. In 1909 the local representative of the British American Timber Corporation at Masset chartered a

Indian Affairs Commission at the Bella Coola Reserve, 1913.
BC Archives H-07096

gas boat and swore in two special constables to patrol the firm's limits to stop Haidas from cutting firewood. But in 1914 the agent discovered a Haida man in a canoe towing a log, allegedly cut on the corporation's claim, up Masset Inlet. He asked the local constable and Indian Agent to issue warnings that any future offenders would be prosecuted. The Indian Agent explained to a superior that both aboriginal and non-aboriginal people combed the banks of Masset Inlet for firewood, and the firm had posted no notices to mark the boundaries of its limits. Going further, he remarked: "Along the shores of Masset Inlet, for miles and miles, is a forest of trees, clear to the foreshore, and no use being made of any of it." The corporation's representative found this willingness to tolerate the traditional practice intolerable, expressed shock at the Indian Agent's defence of aboriginal rights, and reasserted British American's intention to prosecute violators.[9]

While the harsh process of narrowing aboriginal rights to timber, fish and game resources continued to unfold throughout the province, First Peoples at Port Simpson applied to the McKenna-McBride Commission for additional reserve lands. The request included two square-mile plots needed for hunting and timber, both currently held under timber licence tenure. When Metlakatla Indian Agent Charles Perry explained that many of the Port Simpson people were handloggers, seeking to "acquire additional timber lands, so that they may get timber to sell to the mills," the Commission declared that licensed lands could not be considered for reserve extension. Excluded from access to the land base, confined to reserves and lacking capital, aboriginal males would be largely confined to wage labour or at best seasonal handlogging.[10]

Number 2 paper machine, Ocean Falls, 1917.
BC Archives I-47692

Such limited opportunities, at least, began to increase as the industry recovered from the worst effects of the 1913–14 recession. Revival at Ocean Falls came with a late-1914 agreement placing the operation under the control of Pacific Mills, a Crown Willamette subsidiary. The new management secured financing for an ambitious expansion plan, including new sulphite and kraft mills and at least two newsprint machines. "With an illimitable market and an almost illimitable supply of raw material … there is every prospect that the reorganized Ocean Falls Company will become one of the greatest industrial concerns on the Pacific Coast," enthused the *Western Lumberman*.[11]

Engineering and construction work occupied most of 1915, with the *Bella Coola Courier* hailing the prospect of fall and winter work for settlers along the coast. Pacific Mills should give preference to local men, asserted the editor, as "it is from the resources of this area that the company is drawing raw materials." But as full-scale development continued in 1916, manpower shortages prompted the firm to rely heavily on Japanese, Chinese, and East Indian workers. A "Japtown" emerged on the northwest area of the town site that included Chinese and East Indian bunkhouses and mess houses, along with a hall and Buddhist and Sikh temples. Perhaps a thousand men worked shifts of 11 to 13 hours at Ocean Falls by the spring of 1917, while aboriginal and non-aboriginal handloggers built up a log inventory. "The business of logging is very brisk along the neighbouring inlets this season," noted the *Courier*. A number of Bella Coola band members

Dominion Day revelry along 4th Street, Ocean Falls, 1919.
BC Archives I-50641

went to Ocean Falls, assured of handlogging contracts from the company, and non-aboriginal fishermen also sought wage work there after the season.[12]

Newsprint production at Ocean Falls commenced on June 1 1917, and by February 1918 the operation of three machines brought the daily capacity to about 230 tons of paper. The *B.C. Federationist* frowned at the composition of the workforce, described mostly as Italians, Austrians and Germans, "sprinkled with a few other foreign nationalities". The population reached perhaps 2,000 by the spring of 1918, requiring the construction of additional houses, apartments, bunkhouses and schools. A movie theatre was in place by the end of the year, part of the company's effort to attract and hold skilled labour. "The town of Ocean Falls enjoys a picturesque location, and the company is leaving nothing undone that will add to the comfort and happiness of the population," declared the *Western Lumberman*.[13]

Rising wartime pulp prices also contributed to new activity at the Swanson Bay pulp mill, yet to generate a profit for its shifting cast of investors. In 1916 the Empire Pulp and Paper Mills took over the assets of the Swanson Bay Forests, Wood Pulp and Lumber Mills Company as part of an ambitious amalgamation program headed by Ontario lumberman James Whalen. Already in control of the B.C. Sulphite Fibre Company mill at Mill Creek on Howe Sound, and in the process of buying the Colonial Pulp and Paper leases at Port Alice on Vancouver Island, Whalen's group had a crew of millwrights and carpenters refitting the old Swanson Bay

A Japanese crew stacking pulp on the mill docks, at Ocean Falls, November 1922.
BC Archives I-47857

plant by late 1916. Pulp production commenced by the end of the year, with nearly 300 employees turning out 30 to 40 tons of chemical pulp per day. A new shingle mill put the firm's cedar timber to use, and by the summer of 1917, Empire was making regular shipments of pulp and shingles to eastern points along the GTP from Prince Rupert.[14]

Swanson Bay's population reached 500 in 1918 as part of the Whalen Pulp and Paper Mills organization, which sought to raise further capital by offering stock on its three mills and timber assets that summer. By this time the firm had entered the Japanese pulp market, fuelling optimism that inspired further expansion at Swanson Bay. Additional housing for married workers went up, and a larger shingle mill swung into operation early in 1919. The three Whalen plants, including the new Port Alice operation, furnished perhaps one-third of Japan's pulp imports in 1918, employing 1,200 mill workers and several hundred loggers. But the initial returns did not match Whalen's massive investments, and an uncertain postwar pulp market set the stage for yet another reorganization of the troubled enterprise in 1919.[15]

Revival of British Columbia's lumber sector drew momentum from completion of the GTP in 1914, ongoing settlement of the prairie wheat belt and a stronger wartime grain trade. Completion of port facilities in Prince Rupert increased rail traffic, lowering freight rates for lumber moving east. MacMillan also took steps to expand the cargo trade, particularly to the United Kingdom. Shut out of the huge British wartime demand for lumber by American brokers who favoured producers in that country, lumbermen joined the provincial government in pressing Ottawa for help in market

Military officer inspects Sitka Spruce on the Queen Charlotte Islands.
Vancouver Public Library VPL 3868

extension. The pressure resulted in MacMillan's appointment as a special trade commissioner to Britain in 1915. His mission helped convince British officials to increase their purchases of Canadian lumber, including Sitka Spruce for airplane manufacture. Subsequent visits to Europe, South Africa, Australia and New Zealand promoted further expansion of water-borne trade, despite a slide that closed Panama traffic from 1915 to 1917.[16]

Outweighing all of these factors in immediate significance for the north coast, however, was the British War Office's demand for Sitka Spruce, required in the construction of military aircraft. Light and flexible, yet strong enough to withstand the stresses of flight, the species grew in scattered pockets along the coastline. Only on the Queen Charlotte Islands did the extent and density of old-growth spruce timber permit mass production logging, but even here the utilization would be on a selective basis to meet the demand for clear, straight-grain airframe material. Several small mills went into action on the Charlottes by the beginning of 1916, trying to take advantage of the sudden demand for airplane spruce and box material. On Graham Island, plants at Sewell, Port Clements and Masset, all with rudimentary equipment and limited capacity, contracted with Vancouver exporters to handle their cut. When inadequate shipping facilities and high rates cut into both production levels and profit margins, Port Clements operator W.L. Barton arranged with Prince Rupert shipping agents for more frequent stops by GTP, Union Steamship and CPR steamers. But the improved service did not save Barton. Unable to satisfy his creditors, he lost the mill in the summer of 1916. J.B. Weir of Vancouver acquired the property early the following year, making no announcement of future operating plans.[17]

Isolation, lack of infrastructure and competition from well-established Washington and Oregon mills in the British market posed the most serious obstacles to successful spruce production on the Charlottes at the outset of the war. By the time B.C. operators made arrangements to tap the resource there, American lumbermen had secured the first contracts. But strike action by the Industrial Workers of the World slowed production along the American northwest coast in 1917, prompting the Imperial Munitions Board (IMB) to begin looking for alternative sources of the vital war material. The few marginal mills on the Queen Charlottes continued to cut small amounts, and mainland operators also awakened to the increasingly lucrative spruce trade. Whalen's Empire Pulp and Paper Company contracted with the newly incorporated T.A. Kelley Logging Company to commence large scale logging on the Charlottes early in 1917. The logs would be towed to Swanson Bay in Davis rafts, a relatively new technique of securing large assemblages of logs with wire rope to permit transportation across exposed stretches of the Pacific Coast. Although costly to construct and limited to the calmer months, Davis rafting provided the most reliable method of long-distance log transportation until development of the self-dumping log barge after World War II.[18]

Prince Rupert Lumber Company at Seal Cove, ca 1917.
BC Archives A-00437

A major new sawmilling venture at Prince Rupert can also be attributed to the rising value attached to the region's spruce stands. Vancouver
lumbermen J.S. Emerson and E. Duby began negotiating with city council
for a waterfront lease at Seal Cove in late 1916. After securing GTP and city
approval of an 18-acre site for his Prince Rupert Lumber Company mill,
Emerson announced plans for the north coast's largest and most sophisticated sawmill. The two-storey structure would house a 150,000-foot capacity steam and electrically driven mill accompanied by dry kilns, planers
and a shingle mill. An elevated monorail system carried the product
through the manufacturing phases, with a spur line connecting the operation to the GTP tracks. Eventually, Emerson said, deep-water docks for
ocean-going ships would be built for the export trade. Prince Rupert busi-
nessmen expressed confidence that the operation would "be conducted on
a scale that will exercise a pronounced effect upon the general prosperity
of the city and district."[19]

While Emerson proceeded with construction in the spring and summer of 1917, spruce lumbering continued to pick up on the Queen Charlotte Islands. J.B. Weir's Masset Inlet Lumber Company began shipping
clear spruce to Vancouver from the 15,000-foot mill at Port Clements. T.A.
Kelley conducted three camps with 12 large steam donkeys on Lyell Island,
Davis rafting logs to Swanson Bay and Vancouver. Washington and Oregon
producers still commanded the largest share of the export business with
Britain and France, however, until American entry into the war directed
their output to the U.S. government's aircraft program. With this development, the IMB turned its full attention to British Columbia's north coast.[20]

A late-1917 visit by IMB representatives to inspect spruce limits on the
Queen Charlottes, Alice Arm, Swanson Bay and other points proved that

the region's spruce resources were more than adequate, but problems remained. First, private interests held the choicest stands under licence. Second, an industrial and transportation infrastructure had to be erected without delay. H.R. MacMillan had left government to take a management position with the Victoria Lumbering and Manufacturing Company at Chemainus after returning from his trade extension mission, placing new Chief Forester Martin Grainger at the centre of the wartime emergency. In fact, Grainger reported, nothing less than an "industrial migration" was necessary to provide not only for the logging of spruce, but also for the towing, milling and shipping of logs to mills for processing into squares before transportation to Great Britain.[21]

The IMB's Department of Aeronautical Supplies, established in Vancouver that November under director Austin Taylor, assumed responsibility for organizing the spruce drive. Taylor quickly hired MacMillan, who had left Chemainus after a brief and stormy relationship with VL&M's manager E.J. Palmer, as his assistant. Besieged with offers from licence holders anxious to sell their limits to the government, but unhappy with the slow pace of negotiations over the acquisition of suitable tracts, Taylor turned to the provincial government for a remedy. Citing a national emergency, the province took drastic action, adopting an order-in-council that "commandeered" all spruce timber suitable for aircraft components for logging by licence holders. In the case of non-compliance, the Department of Lands received the authority to arrange for logging of such tenures, with the IMB responsible for paying "equitable compensation" for timber cut under these orders. The subsequent Spruce Cutting Act provided for payment at the rate of $6.00 per thousand feet for the highest grade spruce logs, and $2.50 for No. 2 grade. These prices covered waste and damages to timber incidental to the selective logging of material that met the exacting IMB standards. The government exercised its power of expropriation in few cases, and waived its timber-sale procedures to eliminate delays in the logging of spruce on vacant Crown lands.[22]

The letting of contracts began in January 1918, and within a few months the largely undisturbed forests of the Queen Charlottes became the scene of frantic industrial activity. Stand density there permitted the development of large-scale steam logging operations, in addition to handlogging along the shorelines. But predictions of a dramatic expansion of sawmilling on the Charlottes proved unfounded, although some new mills appeared. Smithers businessman W.P. Lynch's Graham Island Spruce and Cedar Company erected a 75,000-foot capacity plant at Port Clements, using machinery taken from a British Canadian Lumber Corporation sawmill in Vancouver. The Masset Timber Company built another larger mill on Masset Inlet, bringing the total there to five by the summer of 1918.[23]

The vast bulk of the spruce cut on Graham and Moresby islands would be sawn on the mainland, at established sawmills. Ocean Falls and Swanson Bay were the main IMB manufacturing centres on the north coast,

Sawmill at Buckley Bay, Masset Inlet, 1918–19.
Vancouver Public Library VPL 3983

although Prince Rupert-area mills also took part. Booms were also Davis rafted south to Vancouver mills by the IMB's fleet of tugs. Riving operations, which involved cutting logs into shorter lengths and splitting these to produce clear squares, allowed transportation by barge across Hecate Strait. The Masset Timber Company became one of the IMB's principal contractors, operating on British American Timber Company limits covering about 100 miles of shoreline around Masset Inlet. By June 1918, the firm employed about 600 loggers, operating two-speed Washington Iron Works yarders described as "the most powerful and the fastest machines yet designed for logging." Elsewhere, Brooks, Scanlon and O'Brien opened a spruce camp on Louise Island. The Abernathy and Lougheed Lumber Company had camps at Skidegate, the Pacific Coast Logging Company operated three camps with 11 donkeys at Cumshewa Inlet, and T.A. Kelly bossed a four-camp, 14-donkey operation on Lyell Island. J.R. Morgan would become another major IMB supplier on the Charlottes, engaging the venerable Georgetown mill to process his airplane spruce.[24]

A shortage of tugs, barges and scows stood as the main barrier to greater production, which reached 30 million feet a month by August 1918. The IMB chartered, leased and purchased the necessary craft, eventually assembling a fleet of twelve boats, five barges, and a large number of scows. Transportation of men and equipment to Prince Rupert and the Charlottes also taxed the GTP's steamship service from Vancouver. Perhaps 1,500 men were situated on Masset Inlet alone, and the presence of several hundred more on Moresby Island prompted the IMB to establish a hospital at Thurston Bay. The IMB's headquarters camp there also included a machine shop, blacksmith shop, warehouses, wireless station and quarters for the loggers. Nearly a hundred logging and riving crews toiled along the entire coast by late 1918 and at least that many steam donkeys had been pressed into service. The operations stretched from Larkum Island, about a hundred miles north of Prince Rupert, where the Granby Company had begun logging its spruce limits, to southern Vancouver Island. Quatsino Sound, Port Renfrew and Nitinat were the primary south-coast production sites, contributing to almost 4.5 million feet of spruce shipped in September.[25]

The evidence is mixed regarding the impact of the spruce drive on seasonal aboriginal activities. One report suggests that many in the Prince Rupert district left the fisheries in the spring of 1918 to work in the spruce camps, attracted by the $6.00 daily wage and abundant overtime. Another later report expressed concern that they would go back fishing as the salmon season picked up, depriving Queen Charlotte spruce operators of their labour. The arrival of several hundred lumberjacks from Quebec and New Brunswick eased the situation, although labour shortages persisted throughout the spruce drive. The YMCA established a combination hotel and recreation hall at Thurston Harbour, the IMB's Moresby Island headquarters, providing movies and reading material for the spruce crews. Another YMCA "hut" was established at Port Clements on Masset Inlet, where the spruce boom had brought the population to perhaps 900 by the autumn of 1918. By that time the dangers associated with rafting logs across Hecate Strait in the fall and winter months had prompted the IMB to sponsor some expansion of milling capacity on the Charlottes, although the report by a Prince Rupert newspaper that mills were "thick on the island" exaggerated the scope of development. By the end of the spruce drive, the Masset Inlet Lumber Company and Sewell Timber and Trading Company operated plants on Masset Inlet. The former, originally the Barton mill, had a 30,000-foot capacity. W.P. Lynch's Graham Island Spruce and Cedar Company's 75,000-foot-capacity mill went into operation at Port Clements at the end of July.[26]

While Queen Charlotte Islands sawmilling gained some impetus from the IMB program, several new plants also came into existence on the adjacent north coast. J.A. McKercher built a new mill at Spruce Bay on Alice Arm in early 1918. J.S. Emerson's Prince Rupert Lumber Company plant went into operation late that year, and Masset Timber Company manager F.L. Buckley established the 60,000-foot capacity National Spruce Mills plant at Kyex on the Skeena River. George Little got into the profitable spruce business at Terrace, and a few miles west at Amesbury a small mill began handling airplane spruce. George McAfee took over the water-powered mill at Georgetown in 1918 to cut squares from spruce logs originating on the Queen Charlottes, and contemplated installing electrical power to make a second shift possible.[27]

Traffic along the GTP line increased as a result of the spruce boom, pleasing officials of the troubled railway. The firm installed two electric cranes at its Prince Rupert dock to move squares from barges and vessels onto cars, which went east to the Atlantic seaboard for shipment to Great Britain. Total spruce output exceeded five million feet in October, and reached 6,850,000 feet in November, 1918. Production may have been even higher had the Spanish-flu epidemic not hit the Queen Charlottes a few weeks before the end of the war. The islands were placed under quarantine on October 15, with the Masset Inlet and Thurston Bay hospitals both filled to capacity.[28]

By the signing of the armistice, Taylor and MacMillan had overseen a truly impressive industrialization of north coast lumbering, especially on the Queen Charlotte Islands. Sixteen logging camps operated on Moresby Island, employing 900 men and more than 50 steam donkeys. Twenty camps were situated on Masset Inlet, providing accommodation for a thousand loggers working in conjunction with 28 donkeys. The last months of the war had also produced further development in sawmill capacity. Two new 60,000-foot capacity mills operated at Thurston Harbour, and another smaller enterprise had been erected 30 miles away at Cumshewa Inlet. Prince Rupert interests had acquired the old Moresby Island Lumber Company mill near Skidegate from the North American Timber Holding Company, established the Queen Charlotte City Mills, and begun turning out 40,000 feet per day at the beginning of October. At Masset Inlet the Graham Island Development Company had put a plant of similar capacity into operation, joining three other small mills, and the Masset Timber Company had expanded its mill to produce 100,000 feet per day with a double shift.[29]

Spruce production totalled over 26 million feet at war's end, "no small factor in winning the war," wrote Martin Grainger in praising industry's "fine spirit of cooperation". But the human sacrifice had been accompanied by tremendous waste in the spruce zones, where low-grade spruce and all the hemlock remained on the ground to rot. The Prince Rupert Board of Trade had expressed concern over the destruction of potential pulp timber on the Queen Charlottes early on, but Grainger considered this a minor consideration "when compared with the saving in time, when time meant a conservation in human lives more precious than any timber." Whatever the economic and ecological cost of the spruce drive, it did not produce the sustained north-coast boom that many hoped for. Minister of Lands and former Prince Rupert resident T.D. Pattullo expressed hope that the spruce initiative would spur long-term population growth. Some Queen Charlotte Islanders had faith that the wartime spruce demand was "certain to lead to big things", perhaps even a rumoured pulp plant for the islands. Pattullo held out the promise of this development and another on the Skeena, "so that the influx of people as a consequence of aeroplane spruce cutting might be kept in the north after the war."[30]

Logged area on the Queen Charlotte Islands, 1918.
Vancouver Public Library VPL 16615

The signing of the armistice left only questions about the future, how-ever, and the initial signs were not promising. On November 20 1918, Tay-lor announced a curtailment of spruce operations, and by the end of the year the owners of three Queen Charlotte Islands spruce mills had disman-tled the facilities for shipment to Vancouver. There was also the question of what to do with the 74 million feet of logged spruce lying in the woods or in booms. In early 1919, Prince Rupert citizens sent a request to IMB and government officials that the remaining logged timber be cut to keep area mills going. A few months later Taylor announced that the logs had been sold, along with all of his department's logging equipment, supplies and tugs. The wartime spruce boom was officially over, then, along with the IMB contracts that had drawn both capital and labour to the region. Whether the artificial demand would leave a legacy of more than sawdust and stumps remained undetermined.[31]

3 The Twenties

Pacific Mills Takes Control

With the important exception of Pacific Mills at Ocean Falls and a growing cedar-pole industry along the Skeena River, the northwest forest industry did anything but boom during the 1920s. The wartime spruce drive had come to an abrupt end, and while some work would be available cleaning up the leftover logs, employment dwindled as most of the small Queen Charlotte Island mills closed. Only a brief period of sawing at Masset Inlet to supply the southern California market with spruce squares interrupted the trend that would draw the Charlottes into the role of fibre supplier to Ocean Falls, Powell River and other mainland paper-making facilities. The Swanson Bay plant consumed some of the Imperial Munitions Board (IMB) logs as well, but failed to survive the decade. A combination of structural problems and corporate decisions doomed it to ghost-town status.

While Swanson Bay withered away, Ocean Falls consolidated its position as the dominant player in up-coast forest exploitation. Cheap and abundant supplies of fibre contributed to Pacific Mills' success in developing a vibrant company town of homes, schools, churches and recreational facilities, one that offered the possibility of family life in exchange for acceptance of paternalistic authority. The firm exercised strict authority over its logging contractors as well, by virtue of its monopoly in the log market. Handloggers and steam operators alike occupied a status akin to southern sharecroppers, beholden to Pacific Mills for financing at the start of the logging season and a market at its end.

If the firm's relationship with suppliers can be described as harsh, that term also captures its relation to the forest ecosystem. The subordination of conservation to profit did not distinguish Pacific Mills from the industry as a whole, of course, a tendency government did little to deter despite warnings from foresters like E.C. Manning. Not until the late 1920s did the B.C. Forest Branch initiate systematic study of the relationship between mechanized clearcutting and forest renewal in the region, perhaps lulled into complacency by a moist climate that made large fires a rarity.

A healthy lumber sector might have ameliorated conditions for contractors and handloggers, but the 1920s brought only frustration for sawmillers. Money might be made on the finest spruce lumber, but north-west forests offered an abundance of low-quality hemlock, balsam and cedar. Competition with lower-coast and central-interior producers in prairie construction markets yielded few rewards on these species, in large part because of an unfavourable freight-rate structure. Manufacturers at Prince Rupert and at points along the Skeena River to Hazelton pleaded for relief to railway officials, but with only periodic satisfaction. H.R. MacMillan led efforts to develop a water-borne trade with Japan but these too fell victim to uncertain demand. Small operators along the Grand Trunk Pacific Railway (GTP) hoped that a large Prince Rupert plant would provide the nexus for shipments through the Panama Canal to the eastern seaboard. That dream faded when J.S. Emerson's Seal Cove plant, under new ownership, burned to the ground late in 1925. Although MacMillan and George McAfee installed a new mill there later in the decade, the entire lumbering sector would be crippled by the Great Depression. Set-tlers and itinerant workers continued cutting railway ties, and Olaf Han-son's new pole business offered another source of employment as far east as Hazelton, but those opportunities too fell off after the October 1929 stock-market crash.

Troubles beset the north coast forest industry in the immediate aftermath of the spruce drive. Although economic conditions seemed stable in 1919, a sharp postwar recession coupled with labour militancy and high freight rates quickly resulted in intermittent operation among lumber mills. With his IMB contracts cancelled, J.S. Emerson leased his new Prince Rupert plant to a group of Vancouver lumbermen who had purchased 15 million feet of spruce logs from the IMB. The mill ran from May until mid December, shipping the cut to prairie and eastern markets. After a shutdown caused by severe weather, production resumed in the new year and contin-ued until exhaustion of the log supply in March. Emerson then resumed control of the north coast's most modern mill, having bought 12 million feet of spruce logs at Masset Inlet from J.R. Morgan. The installation of new gang saws, planers and dry kilns brought daily capacity to 110,000 feet, and optimism ran high that June when Emerson received an order for 1.5 million feet of high-grade Sitka Spruce lumber to be railed to Montreal for shipment to England.[1]

Developments east of Prince Rupert also seemed to herald a postwar lumber boom. Four small mills went into production at Usk in the summer of 1919, one in conjunction with the Kleanza Company's new mining ini-tiative there. George Little's mill at Terrace kept busy that summer cutting for his lumber yards at Smithers and Terrace. The 25,000-foot capacity Lakelse Lumber Company plant at Amesbury resumed operation in May after the winter shutdown. H.R. MacMillan took an interest in George McAfee's Georgetown Spruce and Cedar Company mill at Big Bay, still

Tie camp on the Endako River.
BC Archives NA 03771

relying on water power to turn the machinery. In addition to fish boxes the mill cut spruce lumber for local and rail deliveries, transported by scow to Prince Rupert. Up to 35 small mills operated along the GTP line between the coast city and McBride by mid 1920, ranging in capacity from 10,000 to 40,000 feet per day.[2]

By that time the GTP, unable to pay its bills, had gone into receivership. The federal government, already the unenthusiastic owner of the Canadian Northern Railway, took over the GTP on March 8 1920, creating the Canadian National Railway system. The line faced a growing need for ties to maintain its tracks, giving rise to the "stumpfarm-broadaxe era" in the Buckley Valley. Olaf Hanson, the primary contractor in the region, bought hewn ties from settlers whose pre-emptions held stands of Western Hemlock and Lodgepole Pine. The Swedish immigrant had begun cutting ties for the GTP to finance his Alberta homestead, then moved west to Prince Rupert to secure the contract for the line between the coast and Endako. Wintertime tie-cutting became an integral part of the settlement process along the Skeena during the 1920s, and Hanson ran his own camps as well, purchasing stumpage, laying out roads, and assigning strips to men who received 20 cents for each hand-hewn tie they produced. Hanson would also assume control of cedar-pole production in the region, which commenced in 1919 to supply the North American market with telephone and telegraph poles.[3]

But by the summer of 1920, class conflict and market problems were already apparent along the north coast. The B.C. Federation of Labour had launched an organizing drive in the coastal industry late in 1918, and the

Lumber Workers Industrial Union (LWIU) numbered 11,000 members by late 1919. Briefly affiliated with the One Big Union, the LWIU flexed its muscles, staging 31 strikes in 1919 and another 50 in the following year in an effort to secure higher wages, a shorter workday, better camp conditions and union recognition. The union, with an office in Prince Rupert, mounted its most aggressive effort on the north coast in the spring and summer of 1920. Mill workers at Usk struck on May 1, pressing for increased wages and a reduction in the workday from ten to eight hours. The Royal Lumber Company, Kenny Brothers Lumber Company, Kleanza Company and Hayward Lumber Company all closed down, along with a number of other operations along the GTP line. Loggers on the Queen Charlotte Islands also struck that summer, shutting down the Masset Inlet camps for a time. The Usk operators and their employees eventually settled on a nine-hour day, and District Forester E.C. Manning reported a slight increase in the number of Asian sawmill workers after the conflict.[4]

Increased labour costs for logging and sawmill operators coincided with deteriorating lumber markets. Nowhere was the downturn more evident than on the Queen Charlottes, thanks in part to the cost of transportation across Hecate Strait and freight rate increases imposed by the railway's new management. Several spruce mills there had continued cutting through 1919, handling leftover IMB logs, but by the late summer of 1920 activity had dropped off. Masset Timber Company manager F.L. Buckley denied the evidence of a post-IMB spruce slump, even promising a pulp mill for the islands, but his mill sat idle by October. That left only the Aeroplane Spruce Lumber Company operating on Masset Inlet, and it too shut down shortly thereafter. Manager S.T. Lewis blamed the CNR's freight rate increase for the closure, prohibiting rail shipments to eastern markets. Manning agreed, observing: "The mills on the Queen Charlotte Islands appear to be about out of the rail trade, owing to the freight rates." He advised operators to take advantage of more equitable rates for waterborne shipments to California to sell their spruce in the box market there, and deliver their hemlock logs to the region's pulp plants.[5]

Logging and rafting remained fairly active on the Charlottes in 1919 and 1920 as crews watered IMB logs purchased at bargain prices by the Ocean Falls and Swanson Bay pulp mills and the Prince Rupert Lumber Company. T.A. Kelley contracted to supply Swanson Bay until his firm went into temporary receivership when it was unable to pay employees in 1919. Kelley reorganized, however, and by the spring of 1921 had about 40 loggers cleaning up claims on Cumshewa Inlet. One reporter cited him as the only logging operator on the Charlottes at that time, but the Prince Rupert Lumber Company appears to have had a crew at Masset Inlet before closing its camp in May. That logging and not manufacturing would dominate the Queen Charlotte Islands forest industry became more evident later that summer. First, the Powell River Company, foreseeing the end of its pulp timber on Kingcome Inlet, purchased the rights to over a billion feet of timber on the Charlottes. A short time later the Graham Island Spruce and Cedar Company, the trustee of the Aeroplane Spruce Lumber Company, along with the Sewell and Weir mills, all on Masset Inlet, sold their entire stock of spruce to the Premier Lumber Company of Vancouver. Their inventory, including material still remaining from IMB cutting, would be barged to Vancouver for remanufacturing. "Port Clements, once a hive of industry and the hub of lumber business of the Queen Charlotte Islands ... is now practically a deserted camp," reported a trade journal in the autumn of 1921. The following summer Pacific Mills purchased the rights to 90 square miles of timber on Skidegate Inlet from the North American Timber Holding Company, a $1.5 million transaction that destined the logs for processing at Ocean Falls.[6]

At Prince Rupert, where the collapse of the real-estate boom saw the city taking possession of tax delinquent properties, the lumber industry provided little cheer as the postwar recession set in. George and John Whalen's new Charlotte Islands Spruce Products remanufacturing plant opened on the waterfront in May 1920, resawing spruce squares transported by scow from the Queen Charlottes. But the facility, which included a planing mill, dry kiln and a spur line to the GTP tracks, soon fell victim to falling prices for spruce lumber and the CNR's onerous freight rates. J.S. Emerson's Prince Rupert Lumber Company also shut down in late 1920 after filling a large United Kingdom order. E.C. Manning submitted a bleak year-end appraisal of the north-coast lumber industry's status and prospects. "There is no movement in lumber at present," he reported. "Orders are so lacking that it is useless to quote prices." Factors in the slump included tight credit policies by the banks, high freight rates that had cut off the rail trade to eastern markets except for special orders of clear spruce in long lengths, and low grain prices that reduced buying by prairie farmers. "The industry in northern B.C. is practically at a standstill," reported the *Pacific Coast Lumberman* in January 1921, the Georgetown mill's local fish-box trade providing the only bright spot.[7]

North-coast operators organized a trade association a couple of months later to press for improved transportation facilities and friendlier railway policies. The Northern British Columbia Timbermen's Association blasted new CNR rules limiting the use of hemlock for ties and freight rates that barred shipment of heavy hemlock lumber to the prairies, urging the Dominion Government to build a lumber assembly wharf at Prince Rupert to facilitate cargo shipments from the city. The latter measure might allow marketing of hemlock, the region's most abundant species, on the Atlantic coast. The CNR reversed its decision on hemlock utilization for ties later in 1921, providing a measure of relief for settlers between Prince Rupert and Hazelton, but the industry remained mired in recession. "The inactivity of industry in this district is at present very marked," observed a June report. "Conditions are practically at a standstill and there are no immediate prospects for revival." Only 14 of the district's 42 mills operated that

Buckley Bay wharf with the sawmill behind, 1923.
BC Archives F-05980

autumn, prompting general complaint about inadequate transportation
facilities that hindered northern development.[8]

Manning provided an even more bleak summary of the Prince Rupert
Forest District's lumber industry at the end of 1921. Lumber production
had fallen to 22 million feet from the 1920 cut of 51 million. Tie contracts
amounted to only about half of the 700,000 ordered the previous year, and
the pole industry was "practically dead". Only three or four small mills,
including those at Big Bay and Rivers Inlet had run to fill local demand,
low prices having curtailed shipments outside the region. Although George
McAfee and H.R. MacMillan managed to export almost two million feet
of spruce lumber from Big Bay to eastern North America, England, France
and Australia, prices were so unattractive that many operators "preferred
to keep their lumber rather than accept the losses involved". Wages had
fallen accordingly, from $5.00 to $3.75 per day in interior mills. Pacific
Mills slashed the prices paid its logging contractors, and sparked a strike in
its three company camps by trying to reintroduce the ten-hour day without
offering a wage increase. The firm managed to reopen two of the camps
with new crews, reflecting the LWIU's declining vigour in the face of high
unemployment and a blacklist of union members, part of a more aggressive move toward open shops by employers throughout the province.[9]

On the Queen Charlottes, F.L. Buckley had stirred interest that
autumn by starting the Masset Timber Company's mill to cut spruce logs
that scows moved to Prince Rupert for transfer onto a Canadian Government Merchant Marine vessel just launched from the city's dry dock. The

trial shipment, bound for Australia as part of a cargo of pulp and paper from Ocean Falls, was described as an "epoch making event", perhaps heralding Prince Rupert's arrival as lumber-export centre. But Buckley laid off his night shift after a short period of operation, and shut the plant down completely in November. By the end of the year not a single mill operated on the Queen Charlottes, and practically all of the Masset Timber Company's product remained stacked in its Masset Inlet lumberyard.[10]

Overall, Manning considered the coastal mills in his district to have distinct advantages over those along the government's CNR system. Coastal mills could draw upon better timber, producing a higher proportion of clear lumber demanded in eastern markets. Moreover, they were well positioned to take advantage of cheaper water transportation. But the future for the mainland coast likely lay in the pulp-and-paper field, Manning predicted. As for the interior, crippling freight rates in conjunction with the region's high proportion of low-value hemlock placed sawmill operators at a competitive disadvantage with those further east along the railway. Mills between Prince Rupert and Burns Lake paid 18 cents per thousand feet more for rail shipments to Edmonton than plants east of Prince George, and operators west of town gained no relief in attempting to penetrate the Winnipeg market. Pulp and paper, ties and poles would in all likelihood take precedence over lumber production in the interior, Manning thought.[11]

The recession maintained its grip on north-coast lumbering in 1922, easing somewhat only in the last few months of the year. F.L. Buckley generated one of the only positive developments on the local lumbering scene, amalgamating the Masset Timber Company with the Puget Sound Box Company and a subsidiary of the Los Angeles Drydock and Shipbuilding Company in a scheme to cut spruce squares for remanufacturing into box and construction material in California. Buckley asserted that the new Los Angeles Lumber Products Company's operation would generate 700 to 1000 jobs for the province, a wildly optimistic prediction. Nevertheless, by the end of 1922, 120 workers were occupied at the Buckley Bay mill, supplied by two logging camps. The first steam schooner arrived early in 1923, and reports put the number of workers at 450 that spring, supplying three vessels that ran between Masset Inlet and Los Angeles. The ships carried fuel oil to Vancouver, Anyox, Prince Rupert and other points on the trip north. Port Clements remained a virtual ghost town, however, the government's forest ranger having his choice of 15 rooms in the empty hotel. Then, late in 1923, the Masset Timber Company purchased the old Graham Island Spruce and Cedar mill at Port Clements. By this time the Buckley Bay mill was running double shifts, cutting 260,000 feet a day, and it is doubtful the Port Clements mill operated for any extended period.[12]

To the south, on Moresby Island, several camps operated to supply the mainland pulp-and-paper plants. T.A. Kelley had three camps on Cumshewa Inlet, and the Whalen company conducted logging operations at

T.A. Kelley logging camp at Cumshewa Inlet, 1922.
BC Archives A-05348

Beresford Arm, Thurston Harbour and Sedgewick Bay. Arrangements to facilitate the hiring of loggers at the government's Labour Bureau office in Prince Rupert, rather than the B.C. Loggers Association's hiring hall in Vancouver, drew favourable local comment. On the other side of the employment ledger, Kelley barged logs from his operations on private land to Port Angeles for manufacturing by the Washington Pulp and Paper Corporation.[13]

Along the Skeena River, lumbering remained mired in recession at the beginning of 1923. Thirty per cent of the mills were closed tight, with the remainder "more or less marking time". Operators continued to blame freight rates for their inability to compete in eastern markets, claiming that CNR policies merely deprived the company of traffic. Anxious to develop a water-borne trade, they also kept up pressure on the CNR to build a suitable lumber-assembly wharf at Prince Rupert, and to introduce special "export freight rates" for lumber shipped to the city from interior mills. The Northern B.C. Timbermen's Association passed an early 1923 resolution to this effect, President E. Duby declaring that the key to industrial recovery lay in proper port facilities and entry into the export trade.[14]

Northern logging operators, meanwhile, approached the government for more freedom from Crown land log-export controls. A joint industry-government Export Advisory Committee had been established after the war to rule on applications for permits to ship raw logs to foreign mills, but the government's determination to maintain some control over utilization of the Crown resource made the issue a contentious one. Although mill owners opposed the alternative log market, provincial log exports rose from 10 million to 50 million feet between 1920 and 1922. Much of this material was low-grade cedar, approved for export because of the soft provincial market, and northern loggers desired greater freedom from controls. J.R. Morgan complained that export restrictions were "ripping the logging business in the north", and the Prince Rupert Board of Trade followed up by urging the government to appoint a northern advisory board to consider exports from the region. Until local prices improved, the board requested that approval be granted for Morgan and others to export top-grade cedar and No. 2 spruce logs.[15]

Demand for spruce and hemlock lumber began to pick up by the spring of 1923, providing some hope for a resumption of cutting by idle mills. At that time only the renamed Big Bay Lumber Company mill at Georgetown, the Lakelse Lumber Company at Amsbury, the Kleanza Company at Usk, George Little's Terrace mill, and the Swanson Bay sawmill were running on the mainland, complementing the Masset Timber Company's production on Graham Island. The Royal Mills at Hanall, a 25,000 foot unit founded in 1919, got back into production that spring after two years of sporadic operation.[16]

Olaf Hanson and former District Forester R.E. Allen owned the Royal mill at Hanall, a community of about 100 residents that featured a post office, general store, and school for the children of its 10 families. Hanson and Allen had timber rights sufficient to keep the mill, located next to the CNR track, running for another several years, but faced the prospect of building a railway to haul logs from Pitman and Pacific as they cut out their limits. Terrace was a more bustling lumbering centre. In addition to his main mill located in the heart of the community, George Little partnered with C. Giggey in operating one of several smaller portable plants on the outskirts of Terrace.[17]

Desperate to break free of their dependence on rail shipments, a group of the larger operators along the CNR got together in the spring of 1923 to launch a cooperative effort to enter the water-borne export trade. The Lakelse, Little and Royal operations, assisted by Prince Rupert lumber broker George Nickerson, announced their intention to forward a trial shipment of lumber through the Panama Canal "to advertise the fact that with proper facilities Prince Rupert could develop into an important lumber shipping centre." CNR officials agreed to make certain improvements to the lumber dock, and a Canadian Government Merchant Marine representative pledged his organization's keen interest in promoting the northern export trade. At the same time, the H.R. MacMillan Export Company initiated a project to get northern hemlock into the Japanese market. If prices held, MacMillan said, northern mills should be able to cut sufficient lumber to warrant a trans-Pacific freighter crossing at least every six weeks. MacMillan's enterprise must have been judged the more attractive of the two export options, and on June 27 the Big Bay, Royal, Little and Lakelse mills shipped 400,000 feet of hemlock "baby squares" to Yokohama.

Fume-damaged timber near Anyox.
BC Archives NA-05576

MacMillan's firm negotiated a similar cargo later that summer, and a third before the end of the year. Shipments to the Orient totalled just over three million feet in 1923, and it appears that enthusiasm for the Panama Canal outlet faded in the brief flurry of activity associated with the trans-Pacific trade. The sudden flowering of export business generated further demands for larger and better equipped dock facilities. Lumbermen complained that the existing lumber assembly wharf at Prince Rupert was "proving inadequate and unsuitable for handling of the product".[18]

MacMillan negotiated with other interior plants to contribute to trans-Pacific cargoes, and his Big Bay Lumber Company broke new ground in February by making the first direct shipment of clear spruce from Prince Rupert to the United Kingdom. Logging along the coast also made gains over the previous year, moving into the Alice Arm region for the first time since the war. Tretheway Brothers of Abbotsford took a contract to supply Pacific Mills, built a light logging railway and employed between 50 and 60 loggers. The Granby Consolidated Mining, Smelting and Power Company also began logging a portion of their limits in the Kitsault River valley, extending the Tretheway Railway into the area. Fumes from the smelter at Anyox had killed perhaps 40 million feet of hemlock timber there, and the company decided to reopen its Larcom Island sawmill in hopes of cutting "Jap squares" for the Orient market. Elsewhere in the region the Alice Arm Freighting Company operated a small camp at Silver City getting out logs for the Georgetown mill, C.P. Riel had a small mill at Alice Arm, and Workman and Lawrence ran another at Stewart.[19]

While sawmilling interests struggled to find a stable market, Pacific Mills rose to dominance in the north coast's forest economy, employing up to 400 loggers in its own camps in the vicinity of Ocean Falls. The firm also contracted with J.R. Morgan to cut several million feet from a camp at Surf Inlet on Princess Royal Island. In the Bella Coola Valley, as yet undis-

Granby Consolidated Mining Company Railway, Alice Arm, 1927.
BC Archives NA-05854

turbed by large-scale logging, the Clayton Logging Company contracted to supply four million feet to Pacific Mills. The Bella Coola Logging Company logged a smaller amount of timber destined for Ocean Falls, and the Salloomt Lumber Company ran its water-powered mill intermittently on the logs supplied by local ranchers. Many handloggers and small contractors worked the diminishing number of easy tidewater limits along the coast, bringing the Prince Rupert Forest District's total 1923 log scale to 183,287,904 feet, up about 80 million feet from the 1922 total. The Masset Timber Company's activities on Graham Island, geared to the American market, contributed 25-30 million feet to the 1923 logging output, figuring prominently in the increase.[20]

But Japan proved not to be the saviour for north coast lumbering. The Oriental market continued to generate demand in the early months of 1924, then it slumped. Suspension of shipments forced immediate closure of the Little and Royal mills. Operators along the CNR hoped for a quick resumption of the trade, but MacMillan returned from a visit to Japan with news that no immediate revival could be expected. Although practically all interior mills closed or curtailed production, Olaf Hanson's pole-and-tie industry continued to provide a welcome source of income for workers and settlers along the CNR line between Prince Rupert and Moricetown. While the railway's demand for ties remained relatively stable during the early 1920s, 1924 orders reaching one million, Hanson's cedar-pole business picked up. The district produced over a million feet of poles in 1923 and over 2.5 million the following year, including the output of small operations at Masset Inlet and Bella Coola. In addition to employing over 100 cutters in year-round pole camps at Terrace, Hanall, Pacific, Skeena Crossing and Hazelton, Hanson's wintertime tie camps provided jobs for about 200 tie hackers in 1923. Subcontractors operated other camps, and Hanson continued purchasing poles and ties from area settlers. Despite the arrival

This flume carried water to power the Salloomt Mill at Hagensborg, 1928. BC Archives NA-05542

of Edmonton and Seattle pole firms in 1924, Hanson had achieved virtual control over local production.[21]

Hanson extended the scope of his operations in 1924, using the Skeena as a transportation route for pole drives to assembly yards at Phelan Station, at the mouth of the Skeena, and up-river at Cedarvale. This development opened up cedar stands on tributaries of the Skeena and Kisplox rivers, formerly considered inaccessible. But tie timber within economical hauling distance to the Skeena by horse was becoming incrementally scarce. Stands within five miles of the river had been virtually eliminated by logging or fire, leading District Forester P.S. Bonney to predict the introduction of tractors to cope with longer hauls.[22]

With the Oriental trade dead and access to the North American market uncertain, lumbering remained in the doldrums throughout 1924. An October report described the condition of the district's sawmill industry as "very inactive and unsatisfactory". Hopes that W.F. Buckley's Masset Timber Company would evolve into a sawmilling sector for the Queen Charlotte Islands also lacked foundation. When his Graham Island plant closed in August explanations ranged from a glut in the California lumber market to unspecified management problems. In any event, observers forecast "a long period of idleness" for the 125,000-foot-capacity mill. After disposing of his interests to his Los Angeles partners Buckley bought the Aeroplane Spruce Company mill at Port Clements, demolishing it and transferring some equipment to his adjacent Graham Island Spruce and Cedar Company plant. Buckley announced his intention to begin cutting the following spring, but declared that he would abandon that plan if successful in acquiring the large Prince Rupert sawmill, idle since J.S. Emerson's death in 1921. The Graham Island mill did not run in 1925, and the following May all members of a skeleton staff suddenly left the facility, surprising local interests who had anticipated an immediate start-up. The firm sold its remaining spruce logs to a Vancouver mill for manufacturing there, shipping them south on two log-carrying barges.[23]

By the end of 1924, then, north coast lumbering remained mired in recession. The Prince Rupert Forest District's 19 mills operated well below their total daily capacity of just over one million board feet. Sixteen other portable and semi-portable plants cut sporadically, their operations geared entirely to the requirements of local settlers. Although freight rates had

Cedar poles decked for driving, near Hazelton, 1928.
BC Archives NA-05325

come down to $21.00 per thousand feet from their high of $24.50 on rail shipments to the Atlantic coast, only clear spruce lumber generated a profit. The CNR levied even higher charges on hemlock because of its extreme weight, and the vast majority of mills lacked dry kilns necessary to lighten cargoes. Interior mills, lacking access to Sitka Spruce but with plenty of hemlock, found themselves shut out of continental construction markets. These operations produced only 15 per cent of their capacity in 1924, and their output fell 25 per cent below the 1923 level. Prospects were only marginally brighter for coastal operations. "The timber on the north coast does not yield sufficient high-grade lumber to enable successful operation of large sized lumber mills, except as supplementary operations to pulp-and-paper plants, which could utilize the low-grade logs," District Forester Bonney observed.[24]

Ownership shifts, the most significant involving the Prince Rupert Lumber Company, signalled the arrival of a healthier economic climate by mid-decade. Minnesota interests acquired the Kleanza company's mill and timber limits at Usk and incorporated the Skeena Lumber Company. The 50,000-foot-capacity plant began cutting in January, staffed by about 60 employees. Dealings surrounding the Prince Rupert mill, once the jewel of north-coast lumbering, stirred more local interest because of associated rumours of pulp-and-paper development. The mill and its timber holdings had in fact been the subject of such speculation since Emerson's death. His estate became involved in the Prince Rupert Pulp and Paper Company shortly thereafter, which acquired an estimated billion feet of spruce and hemlock timber north of Prince Rupert from an Iowa firm, and commenced surveys for a proposed pulp plant at Seal Cove. Debt stalled financing of the project, and in 1922 the Prince Rupert Holding Company, an American firm, acquired the paper company's assets.[25]

There the matter rested until late 1924, when F.L. Buckley and Vancouver businessman John A. Smith launched separate initiatives to buy the Seal Cove sawmill. Smith won out, paying an estimated $300,000 for the mill and assets, then established Prince Rupert Spruce Mills and reached an agreement with city officials for tax concessions, power and water supplies. Smith also made vague reference to a future pulp-and-paper project for the city, but even his less grandiose immediate plans for the 150,000-foot-capacity sawmill received a warm reception. The plant would specialize in the manufacture of high-grade airplane spruce stock, Smith said, drawing upon his own timber reserves and those of the neighbouring pulp companies under an arrangement to acquire their No. 1 spruce logs through purchase or exchange. British aircraft manufacturers were eager for his spruce, and he would diversify by establishing a box factory to produce containers for Skeena and Nass River salmon canneries. A new shingle mill would cut cedar for the local construction market. The Seal Cove sawmill resumed cutting on April 1, marking "the commencement of a new era in the northern lumber industry," declared the *British Columbia Lumberman.*[26]

Regional sawmillers cheered the reopening of the Prince Rupert sawmill, hoping that it would provide the basis for them to achieve the long-awaited entrance into the Atlantic coast market through the Panama Canal. Water-borne exports would "be the salvation" of the smaller interior mills, George Little predicted, allowing them to avoid ruinous rail freight rates and contribute to shipments organized by Prince Rupert Spruce Mills. But most of the CNR mills remained closed that spring, the owners hoping for action on ongoing Northern B.C. Timbermen's Association efforts to secure a proper lumber assembly wharf, lower insurance rates on shipping, and establish a brokerage firm to cultivate markets.[27]

Although English orders for high-quality spruce enabled the Prince Rupert mill to cut steadily throughout the summer, and the construction of a grain elevator there generated a short-term demand for construction material, the markets for hemlock, balsam and cedar remained weak. Not even the pulp mills provided a certain outlet for these north-coast trees due to competition from southern producers willing to dispose of their surplus hemlock and balsam logs at low prices. "At present no logger can operate profitably in northern B.C. unless he is getting a considerable percentage of spruce," remarked one analyst. Some relief would come with the establishment of another north-coast pulp mill, a prospect Smith kept alive in press reports. In the interim, the employment he offered 140 mill and construction workers "won the confidence" of Prince Rupert residents.[28]

The north coast's ongoing infatuation with stories of future pulp-and-paper development gained renewed urgency in the wake of the collapse of the Swanson Bay enterprise. A 1919 reorganization, which saw a new board of directors take charge of the Whalen Pulp and Paper Company, had been effected to rescue the Woodfibre, Port Alice, and Swanson Bay operations from a crushing debt load. Former Canadian Pacific Railway executive George Bury had assumed the presidency at that point, supported by a new slate of directors drawn from prominent North American paper making and financial firms. The participation of figures associated with the Mead Corporation of Ohio and the Royal Securities Corporation, both involved in some of Canada's largest eastern paper manufacturers, gave the Whalen company the "experience and financial backing of the strongest interests in the Canadian pulp-and-paper field," declared an industry organ. Royal Securities and a Chicago investment house underwrote $1.5 million worth of bonds, secured by a mortgage on the Whalen assets, providing debt coverage and working capital needed to support Canada's second largest producer of high-grade sulphite pulp.[29]

Bury had severed his connection to the company by the following August, but the initial performance under his tenure seemed promising. New Japanese contracts and more intensive wood utilization from expanded logging operations improved profit margins, and at Swanson Bay cheap spruce logs purchased from the IMB kept the sawmill and pulp plant running to capacity. Introduction of ferry service from Swanson Bay to Prince Rupert cut handling costs, allowing direct transfer of railway cars loaded with lumber and shingles to the GTP tracks for shipment east. The Japanese market absorbed most of the pulp produced at the Whalen mills, production at Swanson Bay reaching an estimated 12,000 tons in 1920.[30]

But despite assurances from new company president T.W. McGarry that the Japanese pulp contracts would keep all Whalen mills running at full capacity for some time, a failing market forced closure of the Swanson Bay mill and logging operations in the spring of 1921. The shutdown, which extended until year's end, probably had as much to do with "bad management as a poor market", E.C. Manning observed. The pulp plant and sawmill reopened early in 1922, shipping products to Prince Rupert for rail transport east. Canadian Robert Dollar steamships also made occasional stops to pick up pulp destined for New York by way of the Panama Canal until dry weather created a water shortage that forced a month-long mill closure in early February. Operations resumed in the autumn, and a May 1923 report indicated that the pulp and lumber mills were running full-time. Activity remained strong that spring, the operation reportedly running "night and day" and "breaking all production records". The car ferry carried nine carloads of pulp, shingles and lumber to Prince Rupert for trans-shipment east by the CNR every week, CPR freighters took pulp south to San Francisco, and Dollar boats loaded spruce lumber for the Japanese market. "The Whalen Pulp and Paper Company is busier at Swanson Bay today than it ever was before," declared a trade journal.[31]

By the late summer of 1923, however, an array of difficulties beset the Swanson Bay plant. Another water shortage caused a temporary shutdown,

Swanson Bay, 1921.
BC Archives I-33734

forcing engineers to cut a new canal from the lake above the mill. That problem, confined to Swanson Bay, paled in comparison to market shifts that caused the entire Whalen organization to collapse under the weight of crippling debts. First, a slump in the price of shingles on the New York market produced significant losses. Even more serious was a massive earthquake in Japan that threatened sales to the market that absorbed roughly 80 per cent of Whalen pulp exports from the Swanson Bay, Port Alice and Woodfibre mills. The collective impact of these events undermined a rickety corporate structure that had only just begun to generate profits after two years of unfavourable market conditions. Unable to pay off its bonded indebtedness from the 1920 reorganization and carry on operations from earnings, the company went into receivership with liabilities of about $10 million after a court application by bond holders. Whalen directors expressed confidence that the firm's affairs would quickly be "readjusted on a satisfactory basis", allowing the mills to take advantage of the anticipated heavy demand for lumber needed to rebuild earthquake-ravaged Tokyo.[32]

But Swanson Bay's problems ran even deeper than published accounts revealed. The firm's capital shortage had prevented needed improvements to the deteriorating facility. Most of the pulp-mill equipment dated from its 1909 installation, and was best suited to handling large logs. The run-down dams, pipeline and buildings would soon require replacement. Although the sawmill had undergone considerable change, it was "neither modern nor efficient as regards its capacity for small timber." Almost half the available timber was too small for the mill to process, according to a government forest ranger. Indeed, the poor quality of timber around Swanson Bay may have posed the most serious obstacle to profitable operation. "A large proportion of the timber in this locality consists of very poor quality cedar," the ranger went on to report. This made profitable operation of the shingle mill difficult, and the hemlock and balsam timber was also lower in quality than that found to the south of Queen Charlotte Sound. Swanson Bay pulp was expensive to produce and prone to discoloration, reducing its market value. Consequently, management had directed large spruce timber suitable for sawlogs to the pulp mill, leaving a good deal of smaller material in the woods.[33]

A court-appointed receiver took authority over the Whalen operations in the autumn of 1923, and the Swanson Bay sawmill apparently operated for a time cutting spruce and hemlock logs on hand. At least one load of spruce lumber went out by freighter to the Atlantic coast, and an order of three million feet of squares for Japan kept some employees at work that winter. A meeting of creditors in Toronto voted to replace the original receiver, an accountant, with an experienced American pulp-and-paper executive. "The situation with regard to the company is very serious," said one bond holder, "and it is only by capable management that the industry will be saved for the province." But the new receiver decided that the plant could not be operated profitably, and the entire operation shut down in December. Only a few employees remained on at Swanson Bay as caretakers, and closure also threatened the Woodfibre and Port Alice plants.[34]

Healthier lumber markets led to the reopening of the Swanson Bay sawmill the following spring, likely cutting existing log stocks boomed at Thurston Harbour. Lumber production continued throughout 1924, freighters stopping to load clear spruce and hemlock for the Atlantic coast market. Although the Port Alice and Woodfibre mills also saw activity, the Swanson Bay pulp plant gathered rust. Then, in the spring of 1925, the bondholders applied for and received an order for the sale of all Whalen assets, valued at $8-10 million. By this time the receiver had leased the Swanson Bay sawmill to the Southern Alberta Lumber Company, which shipped the last cargo of lumber to the Atlantic seaboard at the end of May.[35]

A reorganization plan by bondholders held up the sale until October 30, 1925, when approval of the scheme permitted the creation of the B.C. Pulp and Paper Company to take over the mills, timber holdings and other assets of the Whalen company. The Royal Securities Corporation attributed the Whalen failure to competition from new Canadian mills and exports from Scandinavia that depressed the market for sulphite pulp. Existing logging contracts made wood costs intolerably high, and not until recovery after the 1923 Japanese earthquake did the firm begin to make substantial progress. The almost two-year period of receivership brought further improvement in the organization, allowing operation of the Port Alice and Woodfibre pulp mills. But what would happen to Swanson Bay? Some of the owners apparently favoured continued operation of the sawmill, but initial plans did not include reviving the north coast facility.[36]

Healthy pulp markets permitted steady operation at Port Alice and Woodfibre in 1926 but Swanson Bay remained idle, its plant "rapidly deteriorating". Rumours circulated that B.C. Pulp and Paper would modernize, or that development might take the form of a rayon silk facility. But as time passed, and the entire community fell into disrepair, even optimists conceded that massive reconstruction would be required to resume production. B.C. Pulp and Paper retained possession of the original pulp leases, and the Swanson Bay area remained a site of logging for the mill at Ocean Falls. Pacific Mills contractor Mark Smaby logged over 15 million feet around Swanson Bay in 1928, and handloggers continued to live there. In 1940, 42 people called Swanson Bay home, including a B.C. Pulp and Paper caretaker. The population had dwindled to 12 by 1943, the final year Swanson Bay appeared in the B.C. Directory. The community presented a scene of complete desolation when Bill Moore made a visit in the 1960s. A piece of the smokestack remained erect, the mill walls had crumbled, and "the shingled roof of the homes lay folded upon the ground like a collapsed deck of cards, with large alder trees growing up through them."[37]

If Swanson Bay represented a conspicuous failure of the provincial government's early 20th-century pulp-and-paper initiative, Ocean Falls conveyed a more positive example of business-government cooperation. During the 1920s, Pacific Mills succeeded in developing a community that embodied both the advantages and shortcomings of British Columbia's single-industry towns. For resource workers, communities like Ocean Falls offered the opportunity for a stable family life, albeit one in a damp, isolated and increasingly polluted setting. Another price would be subjection to the heavy hand of corporate paternalism. Firms such as Pacific Mills exerted an ever-present influence on workplace and social life, blending benevolence and authoritarianism in its relations with the inhabitants of Ocean Falls.

The community's dismal climate and isolation from major population centres encouraged Pacific Mills officials to devote considerable attention to labour stability, especially among skilled paper makers. While the "bunkhouse man" remained a significant element in the unskilled labour supply, the provision of single-family residences, duplexes, schools, playgrounds, churches and recreational facilities reflected the firm's need to attract and retain workers. Beyond that operational imperative, executives no doubt shared the common perception that family life encouraged the creation of a more docile labour force. Not only did such ties create dependence upon a particular company, reducing the chronic turnover that undermined efficiency on the industrial frontier, some thought that wives would exert a calming influence when strikes threatened family income. But like most Canadian company towns of this era, Ocean Falls did not reflect the sophisticated town-planning philosophies that influenced the design of some American inter-war industrial settlements. The terrain at the head of Cousins Inlet offered little space for innovation, dictating that

Japanese bunkhouse at Ocean Falls, 1925.
BC Archives I-50621

the town site would develop in close proximity to the mill. Streets were laid out to maximize the very limited level terrain, with some wooden avenues reaching up the hillside to increase the residential area. Occupational and racial segregation also influenced the configuration of housing. Highly skilled papermakers and other technical staff congregated along with supervisory personnel in the largest houses along Front Street, while Asian workers occupied their own community, "Japtown", separated from the town site.[38]

Over the years, Pacific Mills succeeded in cultivating a stable core of skilled workers, awarding service pins to loyal employees at annual company dinners. Turnover remained high among unmarried labourers, however, keeping the firm's Vancouver employment agent busy when mill expansion and healthy paper markets created peak labour demand. During the construction phase, for example, the agent advertised in the city's skid-row bars, rounding up his hires the next morning at the dock for departure by coastal steamer. The greatest problem, he recalled, was getting his hung-over crew to board ship for the voyage to the damp, isolated community. Ocean Falls might offer workers more attractions than the typical coastal logging camp or sawmill village, including a nearby brothel, but for many it would represent just another resource-exploitation site where one could accumulate a "stake" before returning to Vancouver. "It was a good place to work and raise a family," recalled Ocean Falls resident Roy Nakagawa, "but it was a boring place for single working people who lived in the bunkhouses."[39]

"Boat Day" at Ocean Falls, 1917.
BC Archives I-50600

For those tied to the harsh treadmill of resource-camp life, however, family housing and corporate paternalism conferred very real improvements in living standards. When Howard Phillips stepped onto the Ocean Falls dock with his father in 1919, his senses were assaulted by "the sulphate smell, ... the sound of the mill in action, and the sound of the water over the dam." The distinctive wooden streets caught his eye, and he found the housing conditions "utopian". The family's company house featured steam heat, electric lights and indoor plumbing, all new to Phillips, and Pacific Mills distributed a free ham or chicken to families every month. None of the residents owned automobiles, limiting traffic to company vehicles, including a truck that delivered groceries. The arrival of a steamship bearing foodstuffs, catalogue orders and visitors brought residents to the dock, making "boat days" welcome events at Ocean Falls and other coastal communities. Such stops broke the routine and provided the opportunity for an occasional trip "outside".[40]

But British Columbia's company towns had a darker side as well, existing outside the framework of municipal laws that regulated urban settlements. No elected council existed to moderate the company's absolute control. Firms such as Pacific Mills owned the land, housing, mill, store and recreational facilities; they controlled access to their communities, and set prices at the company store, rental charges for accommodations and rates for utilities. "Everything belongs to the company," observed the *B.C. Federationist* in a critique of these feudal arrangements. "The worker lands on a company wharf, eats at a company hash house, sleeps in a company bunkhouse, buys his booze at a company bar, and his 'terbacker' and over-

Ocean Falls
town and mill, 1926.
BC Archives I-48458

alls at a company store. If he dares to make a kick, he is escorted from the company premises by a company constable."[41]

The provincial government made a half-hearted effort to increase rights of access to these settlements in 1917, responding to mounting opposition to the autocratic control exerted by owners who routinely sent union organizers and other undesirables packing. Liberal Minister of Lands T.D. Pattullo proposed legislation that year that would have compelled corporations to register town plans with the government and convey to the province a one-quarter interest in town lots. Corporate objections in conjunction with the dubious legality of the measure, which posed no fundamental challenge to owners' rights, prevented its implementation. Two years later Pacific Mills flexed its muscles by sending two union organizers back to Vancouver right after learning of their activities. The firm might have been a "benevolent and generous autocrat", official historian J.S. Marshall later wrote, but an autocrat it was.[42]

By the end of 1919, the Ocean Falls mill operated around the clock, seven days a week, to keep up with newsprint demand. At the dock, workers loaded freighters bound for San Francisco, Seattle, Australia and New Zealand. As output remained high throughout 1920, town site development continued with the addition of family homes and bunkhouses. The hotel lobby, site of a barbershop, cigar stand and pool tables, became a

The Ocean Falls swimming pool, 1929.
BC Archives I-48942

popular male gathering place. The end of prohibition in 1922 brought a
government liquor store to Ocean Falls, but residents voted to close it in
1926 after a series of waterfront drownings. Pacific Mills then installed a
beer parlour in the hotel, devoting profits to town site improvements.
Anglican and United churches provided a countervailing influence. Revenues from the pub went into the construction of an indoor swimming pool
in 1928, ensuring that the community's children would learn how to swim.
Considered vital to the safety of youngsters given the town's waterfront
location, the pool would go on to serve as the home of a swim club that
produced British Empire, Pan-American and Olympic Games medalists. A
baseball park and playing field for soccer and lacrosse competitions provided further opportunities for organized sport. Ocean Falls possessed "all
the requirements of an up-to-date Canadian city", enthused a trade journal.
The pace of production slowed as the recession of the early 1920s set in,
bringing 1921 wage cuts and temporary closures of some departments.
Although the mill operated consistently after the economy revived in 1923,
overproduction in the North American newsprint market narrowed profit
margins throughout the 1920s.[43]

The 1924 demographic makeup of Ocean Falls reflected the transition
from frontier work camp to industrial community. A total of 997 men, 416
women and 513 young people (under the age of 21) brought the population
to 1,926. The young people consisted of 262 children under 6 years of age,
213 from 6 to 16 years old, and 38 from 17 to 21. Caucasians dominated the
town's ethnic structure, numbering 1,303. They lived in close proximity to

but largely independent of 538 Japanese, 59 Chinese and 26 East Indians. Restrictive immigration policies ensured that the Asian community lacked the age and gender balance evident among whites. Of the 416 women who resided in Ocean Falls, 328 were Caucasian. Only among the Japanese was a semblance of family life achieved, marked by the presence of 88 Japanese women and 114 children. No Chinese or East Indian women and children appeared on the census rolls, thanks to the province's heritage of anti-Asian sentiment that isolated these male resource workers from families in their homeland.[44]

Logs for the Pacific Mills plants came both from company camps and a "colourful assortment" of handloggers and contractors. Company operations at Green Bay, Kimsquit and Kwatna employed over 300 men, utilizing sophisticated overhead logging technologies linked to industrial railways that extended to tidewater. The handloggers took much smaller amounts of timber from the inlets and bays, and logged the shores of Link Lake and its rivers. But by the early 1920s, Pacific Mills was having difficulty meeting its timber needs from its leases, amounting to two billion feet annually. A 1914 amendment to the province's Timber Royalty Act had extended the pulp companies' leases for 30 years at the original low royalty rates, and they had also been aggressive in taking up timber sales in an effort to monopolize local supplies. These too offered lower royalty rates on pulpwood than sawtimber, an advantage Pacific Mills executives defended on the grounds that pulp or paper manufacturing generated more employment for Caucasian labour than lumber production did.[45]

Foresters acknowledged the validity of this claim, but rejected another corporate justification for low royalties based on their supposedly more complete utilization of the resource. While true in theory, inspections showed that hemlock utilization by "ordinary loggers" on timber sales equalled that of the pulp companies. Moreover, the coastal pulp mills used high-grade spruce logs desired by lumber producers. Finally, pulp executives such as Pacific Mills' Archie Martin argued that the public benefits derived from north-coast industrial development entitled these firms to cheap timber. But this timber was a public asset guaranteed to rise in value, a forester pointed out, one which should not be "forced upon the market prematurely and too cheap". And did these enterprises actually *require* stumpage concessions? The original leases had promoted more speculation than development. Did the logic still hold that low royalty rates fixed far into the future were essential to attract investment capital? The 1914 Royalty Act had introduced a sliding scale of royalties for the sawmill industry, based on the selling price of lumber. The same approach should provide sufficient tenure security for the pulp manufacturers, one analysis suggested. "Why western pulp mills should not be in a position to pay for timber just as sawmills do I cannot see," the author argued. Firms such as Pacific Mills already possessed "large valuable concessions which cannot be duplicated, and they should thus be all the less in need of further bounties."[46]

Spongberg, a logging contractor, had his logging camp at Gardner Inlet in 1921.
BC Archives I-33737

This sort of thinking exerted some influence on 1921 legislation that met the demands of pulp companies for exclusive rights to further timber supplies. The measure gave the Minister of Lands the authority to establish pulp reserves in the area of existing holdings, free of annual rental charges. The pulp manufacturers could draw on these as they wished, but would pay the stumpage rates charged on timber sales at the time of cutting. Pacific Mills immediately secured the rights to Pulp District One, gaining the privilege of cutting up to a billion feet of timber from the area. A few years later the firm would be back in Victoria, complaining about the quality of timber in the reserve. Accordingly, Pacific Mills received the rights to yet another reserve of 300 million feet, Pulp District Two, in the Swanson Bay area. Minister of Lands T.D. Pattullo made this award, which brought its holdings to at least 3 billion feet of timber, conditional upon the company making no further application for unalienated Crown timber. During this period Pacific Mills had extended its reach over much of the north coast, drawing timber from its original pulp leases, the new reserves and its Queen Charlotte Islands holdings where contractor J.R. Morgan operated. At Alice Arm the Tretheway Brothers conducted a large operation, and other contractors worked Gardner Canal, Kitimat Arm and Swanson Bay. Still not satisfied, by the end of the decade the firm requested that Victoria grant it the rights to another 100 million feet beyond its pulp districts.[47]

During the early 1920s, Pacific Mills turned its company camps over to contractors, and in 1923 Mark Smaby took on the responsibility of overseeing all of the handlogging and contractor operations. From the start, the firm emphasized immediate profit over long-term sustainability in developing logging plans, "picking out the best patches of timber along the face of their pulp leases" and leaving less-accessible timber for the future. E.C.

Another camp at Labouchere Channel, 1927, run by a contractor named Soderman.
BC Archives I-48686

Manning expressed concern over the Forest Service's inability to regulate cutting on the original leases, given the favourable royalty rates on pulp timber. The agency had stronger authority over utilization practices on timber sales, but Manning predicted that the company's "short sighted" approach to its leases would have serious consequences.[48]

With the demise of the Swanson Bay plant, Pacific Mills assumed virtually absolute control over north-coast logging. "The logging industry on the whole mainland coast is gradually coming under the control of the Pacific Mills," Manning reported at the end of 1923. Even when Swanson Bay operated, the two firms had not competed for logs, fixing prices "on a strictly supply and demand basis regardless of whether the logger gets equitable returns or not." As the sole major log buyer after 1924, Pacific Mills drove hard bargains with its handloggers and contractors. Smaby and his successor Parker Bonney, who succeeded Manning as District Forester before taking corporate employ, "saw to it that none of the contractors for Pacific Mills made any real money", writes James Sirois. Moreover, the firm paid a standard price for pulpwood logs, then sorted out the best and sold these at a profit on the Vancouver market. After paying off the money advanced by the company to purchase supplies and equipment, the logger "often had damn little to show for the wear and tear on his hide. He was lucky if he made wages." According to some accounts, the Haisla handloggers at Kitimat suffered a sharp drop in income after closure of the Swanson Bay mill terminated that log market and placed them at the mercy of Pacific Mills.[49]

On the north coast, then, the quasi-monopoly position enjoyed by Pacific Mills gave the firm enormous power over its log suppliers. "Reaching out for all the available Crown timber which constitutes fair logging shows," the region's dominant entity strove to control both the resource and the market to assure itself of cheap raw material. And in an area where easily accessible, high-quality timber was at a premium, the existing pattern of corporate holdings in conjunction with an aggressive pace of exploitation left little room for the emergence of an independent logging sector. "The waterfront shows are rapidly being cut out," Manning observed as early as 1923, "which will mean going back farther in larger operations." Around Swanson Bay too, the most accessible timber was locked up in pulp leases. "There is no reason why independent loggers should come to the locality," the ranger there observed.[50]

Under such circumstances the Forest Service struggled to introduce reasonable standards of forest practice on timber sale operators whose profit margins allowed little consideration for utilization standards and resource renewal. Although agency resources were stretched too thin on the north coast to allow systematic silvicultural studies, Manning was pessimistic about the rate of regeneration on the larger clearcut areas. "In the smaller donkey operations where the logs are simply 'siwashed' in, all the remaining young timber and seed trees are not destroyed and satisfactory growth will probably take place," he reported. "But in the larger donkey operations, the little evidence to hand does not indicate that satisfactory reproduction is coming in." More positive conditions prevailed in the interior, where selective horse logging of cedar and pine for ties and poles left undamaged residual timber as seed sources, but in neither area had sample plots to monitor reproduction been established.[51]

Much of the coastal logging during the 1920s and subsequent decades involved A-framing, an ingenious adaptation of technology to steep shoreline slopes. Operators placed their steam donkeys on log rafts stationed next to logging sites, extending the mainline through a block atop the frame up the adjacent hillside. Logs could then be yarded directly into the water for booming, or "swung" from cold decks accumulated by another donkey. A bunkhouse, cookhouse and machine shop might be placed on a nearby float, permitting movement of the entire operation along coastal inlets. Only the availability of a fresh water supply limited the location of these floating camps, James Sirois recalls, accessed by damning the creek at a point high enough to generate sufficient pressure to carry the water through a line to the camp.[52]

P.S. Bonney's 1924 annual report on the Prince Rupert Forest District expressed another dismal appraisal of industry forest practice, described as "appallingly crude and wasteful". Problems in utilization, slash disposal and regeneration would "ultimately prove very heavy handicaps to future administrations, [but were] impossible of immediate solution in the face of existing economic barriers," Bonney observed. Bonney's staff had begun taking action in timber sale administration on the coast, reversing a practice that had allowed small operators to high-grade "cream chances" near shorelines. The new policy would limit sales to complete logging units,

A-frame logging show, 1949.
BC Archives NA-12098

eliminating "considerable butchery", but perhaps making life even more difficult for the already oppressed small contractor. Bonney repeated Manning's plea for research to serve as a basis for future silvicultural requirements. "The time is fast approaching when a beginning must be made in at least the primary features of forest management," he advised.[53]

Forest Service reproduction studies on the coast did not extend past Ocean Falls at this time, although a survey of cutover areas in the hemlock-cedar forests from Cracroft Island to that region seemed to indicate plentiful natural reforestation. Not until 1927 did the agency commence a similar study on the north coast, surveying cutovers on the Queen Charlotte Islands and around Ocean Falls logged during the World War I spruce drive. Sample plots revealed that 80 per cent of the areas contained over 500 seedlings per acre, with hemlock slightly more prolific than cedar and spruce in the new forest. The primary pulp species of spruce and hemlock together constituted 69 per cent of the reproduction in the region, positive findings given the predominance of pulp-and-paper production in forest utilization. Generally undisturbed by fire because of the moist climate, north coast forests exhibited a more satisfactory rate of natural renewal than the Douglas-fir areas on the south coast. There, alarms were already being raised about the massive scale of clearcuts and the repeated fires that swept through slash-filled cutovers.[54]

Such rudimentary research results would have done nothing to encourage the introduction of more rigorous forest management standards along the north coast. In any case, most foresters still considered fire a more serious threat to forest renewal than cutting practices, although E.C. Manning would campaign for clearcutting regulations during his tenure as Chief Forester during the late 1930s. High annual precipitation and cool temperatures generally made fire in the Prince Rupert Forest District a less pressing concern than in drier regions, but did not guarantee immunity. Over

70,000 acres of merchantable timber had been destroyed or damaged by fire during the very dry spring and summer of 1922. Moreover, the scattered population and difficulty of organizing and maintaining fire crews in an area with a poorly developed transportation infrastructure continued to deter efficient detection and suppression. Thus, the agency had difficulty responding when abnormally dry weather beset the eastern part of the district in 1930. Blazes damaged over 96,000 acres of timber that fire season, 51 per cent of the area burned over in the entire province.[55]

Although prices remained volatile, the provincial lumber industry boomed during the second half of the 1920s. The value of production hovered around the $40 million mark from 1924 to 1927, hit a new record of over $48 million in 1928, and climbed to $50 million in 1929. But north-coast lumbering continued to be beset by its historic problems: competition from southern and eastern mills, high freight rates, and difficulty in marketing anything but high-grade lumber at a consistent profit. Making matters worse, a fire in November 1925 wiped out much of the new Prince Rupert Spruce Mills plant. The loss, estimated at $500,000, involved complete destruction of a new planing mill and box factory, along with a good deal of lumber stored in the yard. One commentator described the event, which threw 75 men out of work, as "probably the worst blow that has ever befallen the lumber industry of Northern British Columbia." Smith conceded that the plant had not been fully insured, and warned that reconstruction would have to await the decision of insurance adjusters.[56]

The fire at Seal Cove contributed to a bleak outlook for an industry that had struggled all year with the usual conjunction of low prices and onerous freight rates. Mills had found it difficult to break even, outgoing Northern B.C. Timbermen's Association president Olaf Hanson said at his address to the annual meeting in January, 1926, and no progress had been made in freight rate equalization, royalty reduction or market extension. But the new grain terminal might allow lumbermen to place hemlock and cedar squares on vessels loading grain for the Orient. A local brokerage firm would also assist in developing profitable outlets. But the association would have to get along without John A. Smith and the Seal Cove mill. His health broken, Smith lay in a California sanatorium that winter, later committing suicide. H.R. MacMillan and George McAfee then purchased the site from the CNR, announced their intention to move their Big Bay Lumber Company mill at Georgetown to Prince Rupert and have an expanded plant in operation by the summer of 1927.[57]

MacMillan and McAfee began clearing at Seal Cove while negotiating for tax concessions with city officials. Their Georgetown mill continued cutting for the local cannery and halibut box trade that summer, described as an average one for the industry. George Little and some associates acquired the Lakelse Lumber Company mill at Amsbury and began cutting after two years of inactivity, and Little's Terrace plant ran continuously. But a couple of other CNR mills suffered severe setbacks. The Skeena Lumber

Company operation at Usk went into bankruptcy, and the Canada Products facility near that community was lost to fire. On the Queen Charlotte
Islands, F.L. Buckley took over the old Lynch mill at Port Clements that
spring, refitted the facility, and cut some spruce until Christmas. At Queen
Charlotte City, the Southern Alberta Lumber Company operated the Sitka
Spruce Mills plant from 1925 until shutting down late in 1926 because of
financial problems.[58]

Of the 20 permanent mills in the Prince Rupert Forest District, 13 operated at some point during the year. Another 13 portable mills cut "spas-
modically", but all "suffered from lack of profitable market". Foresters
attributed this "depression" to overproduction of the lower grades of lumber produced by the region's mills Operators simply could not compete
with southern mills which tended to "sluff this grade off for whatever is
offered for it, figuring on getting their profit out of the higher grades of
which there is much larger percentage than obtainable from the timber cut
in this district." Those producers also continued to have the advantage of
cheaper ocean freight rates to eastern markets. Forest Service staff agreed
that "the mill men had a very hard time, several closing down, others going
into liquidation." The few that continued to operate did so under adverse
conditions, with the result that "nobody made any money and most of
them lost money."[59]

The hewn-tie industry, which by one conservative estimate produced
over six million ties in the area between Prince Rupert and Endako from
1919 to 1930, entered decline of a more permanent sort. Production in the
Prince Rupert district began falling after 1924, when over a million ties
were shipped. The 1926 output totalled only about half that amount, and
the opening of a creosoting plant in Edmonton the following year dashed
hopes for a revival. The creosoting process had important consequences
for tie hackers. First, treated ties had three times the life span of untreated
ties. Second, the process demanded that ties be peeled, a time-consuming
task that the CNR did not reward by increasing prices enough to prevent
many from leaving the industry. Finally, creosoting encouraged the production of mill-sawn ties, since the surface accepted the preservative as well
as the smooth finish produced by hewing. The tie industry had "passed its
zenith", reported the *Prince George Citizen* in 1929. "Treatment of the ties
with creosote has prolonged their life, and the orders of the railway company for ties are being curtailed each year."[60]

More prolific pole production in the district offset the dwindling tie
business to some extent. Hanson's operations along the Skeena and Kispiox rivers coincided with the drive to Cedarvale, where up to 1,000 rail cars
were loaded for shipment east, in 1927. Output rose by over 30 per cent the
following year, thanks in part to J.H. Baxter's operation at Masset Inlet.
Baxter had gone into business there in 1925, exporting poles to California
from excellent cedar stands at Ferguson Bay. In 1928 he began offering
contracts to local operators to extract poles from Crown grants in the

Pole drive on the Skeena River, 1928.
BC Archives NA-05570

Mayer Lake vicinity, building a four-mile road to Kundis Slough upon which he planned to operate a tractor and trailer equipped with concave wheels to run over hemlock rails.[61]

Sawmilling activity in the Prince Rupert district remained minimal, however, reflecting a slump in lumber values that troubled operators around the province. MacMillan and McAfee finally began construction of their promised mill at Seal Cove in the summer of 1927. Their Georgetown mill continued to operate steadily, but of the CNR mills only George Little at Terrace, now logging with at least one Fordson tractor, maintained a similar basis of production. Things were even quieter on the Queen Charlottes, where bankruptcy claimed the Canada Lumber Yards mill at Port Clements and the Sitka Spruce Mills operation at Queen Charlotte City. Toward the end of 1927 a rumoured reopening of the Masset Timber Company mill at Buckley Bay sparked interest, but its California owners announced that they would await improved market conditions. The installation of milling machinery at Seal Cove by the Big Bay interests that October, with production expected to commence the following March, provided the only heartening news in a year devoid of major developments.[62]

Coastal logging contractors also experienced hard times as the overproduction of logs on the south coast gave the pulp mills a cheap alternative fibre supply. Unable to secure another contract with the Powell River Company, T.A. Kelley shut down his Queen Charlotte Islands camp in May after ten years of operation. That threw 200 loggers out of work temporarily, but the islands figured prominently in the long-term operational plans of the paper makers. The Powell River Company negotiated the purchase of a billion feet of pulp timber on Graham Island from the province early in 1927. Northerners expressed regret that the timber would not be manufactured locally, but took some consolation in knowing that it would not find its way to American mills.[63]

But as the firm began surveys of a logging railway to Skidegate Inlet, some islanders took issue with the locking up of the resource for future exploitation by giant enterprise. A Queen Charlotte City resident com-

Loading logs on a Blue Ox logging truck near Terrace, 1928.
BC Archives E-01794

plained to new Conservative Minister of Lands F.P. Burden that his Liberal predecessor, T.D. Pattullo, had "tied up all available timber nearby with large outfits and practically told us small operators and handloggers to move off the Island." The Skidegate Inlet Conservative Association asked Burden to make no further timber sales unless accompanied by a clause calling for immediate utilization. "The large Companies like the Ocean Falls, the Powell River and the North American Timber Holding Company have large holdings at present, especially around Skidegate Inlet, which they are holding for future use." the members pointed out, "making it very difficult for the present population to make a living."[64]

A general improvement in lumber prices during 1928 set the stage for the long hoped-for expansion of sawmilling capacity around Prince Rupert. MacMillan and McAfee spent the early part of the year moving equipment from Georgetown to the new 50,000-foot capacity Seal Cove mill. The venerable Georgetown site went silent for a time that spring, an event that drew only passing comment in the excitement that accompanied the April opening of the new Big Bay Lumber Company mill on the eastern side of Prince Rupert. "The industrial rhythm of the saws once again pervades the air at Seal Cove," the *British Columbia Lumberman* observed, making "welcome music to local ears." The plant, which promised 60 full-time jobs, shipped clear spruce east on the CNR in addition to cutting for box shooks for the fish trade. Louis Locker supplied the company with logs, cutting Big Bay's limits on Porcher Island before moving on to timber sales and licences at Baker Inlet. Manager McAfee also purchased high-grade spruce logs from J.R. Morgan, and arranged for Pacific Mills to take pulp-quality material.[65]

Morgan, one of the district's major logging contractors, also made an effort to get into the sawmilling game after finishing up his contract with Pacific Mills at Surf Inlet late in 1927. Morgan then moved his 75 loggers to Mussel Inlet to fill a contract with Pacific Mills for 10 million feet of logs. About 35 miles north at Swanson Bay Mark Smaby supplied over 15 million feet to Ocean Falls, subcontracting to eight steam-powered operators and a number of handloggers. Outweighing the constant shifting of the north-coast loggers in importance, however, was Morgan's announcement that he had rounded up a group of investors to establish a sawmill just south of Prince Rupert at Porpoise Harbour.[66]

Morgan and his associates broke ground at the Billmor Spruce Mills site in May 1928, planning a 40,000-foot capacity plant capable of participating in the fish-box business as well as exporting via either ship or rail. Local commentators anticipated an autumn start-up, but the completed plant sat idle at the end of year. It remained quiet through the following spring, as Morgan moved his logging operation from Mussel Inlet across Hecate Straight to Skidegate Inlet. His high-quality spruce logs would go to the Billmor plant, the operator announced, with the pulp logs destined for Ocean Falls. But at the end of 1929 Morgan set the following spring for the opening, a plan that the Great Depression would inevitably disrupt.[67]

A similar fate befell the logging and sawmilling venture launched by the American holders of extensive limits on the Ecstall River. There James Brown continued to cut steadily, supplying box material to the canneries, but R. Armstrong of New York and J. Randall Black of Illinois had more ambitious plans for the area. Their National Airplane Lumber Company went into action in the summer of 1928, using a gas-powered donkey to yard clear spruce logs to Big Falls Creek for driving down to Big Falls. There they would be yarded to a chute that extended down the bluff into the Ecstall River, which would carry the logs into the Skeena near Port Essington. Their destination would be a new 40,000-foot capacity mill at Seal Cove, near the Big Bay Lumber Company plant. The choice spruce would be cut into airplane stock for shipment east by water or rail. But finding that site too small, they went ahead at Porpoise Harbour instead, on ground adjacent to Morgan's Billmor mill. Construction on a small steam plant began at the beginning of 1929 in preparation for a larger facility, but at some point that year the project stalled. District Forester R.E. Allen reported at the end of 1929 that construction had halted, and the logging operations at Falls River appeared "permanently abandoned".[68]

As the north coast drifted toward depression in 1929, then, only marginal gains had been made in mainland sawmilling. Logging contractors and handloggers remained subject to the dominance of Pacific Mills, which increased the number of contracts in 1929 after depleting much of its watered log supply the previous year. Contractors in the Swanson Bay district supplied 21 million feet, and the firm drew 57 million feet from Pulp District No. 1. A portion of the log supply also came from the Queen Charlotte Islands, where the logging industry enjoyed a brief pre-depression boom. After commencing operations on Masset Inlet in 1928, the Allison Logging Company moved to Moore and Inskip Channels, establishing a headquarters camp and loading works on a large barge to permit flexibility in moving the outfit. A.P. Allison also attempted the first commercial logging on the west coast of Moresby Island, a single A-frame operation and several handloggers working around Douglas Harbour and Peel Inlet. He employed barges to ship most of his 1929 output of 14 million feet to Ocean Falls and Powell River, selling the higher-grade spruce and cedar to Vancouver lumber mills. Elsewhere on Moresby Island, T.A. Kelley got back in business at Selwyn Inlet, building a short truck road for hauling logs produced by four donkeys and 65 loggers. He rafted most of the 10 million feet cut to the Big Bay mill at Prince Rupert. J.R. Morgan operated at Skidegate Inlet on Graham Island, utilizing three donkeys to log First Nations reserve timber and a small government sale.[69]

Market conditions allowed profitable operation of interior mills during the first half of 1929, then orders vanished as demand slackened. Settlers produced about a million ties for the CNR, the company giving preference to local residents in awarding small contracts. The railway company's orders would continue to decline, but R.E. Allen praised the policy of favouring the "local resident public" over transient labourers "who either take their stakes out of the country or spend a large percentage of them in places of amusement." More of the settlers' income tended to remain in the region, passing into the hands of local merchants, than that of the "tie-hack-harvest-hand who spends the winter time in the woods and migrates to the harvest fields after having spent the greater part of his stake in the beer parlours and pool halls."[70]

Allen expressed real optimism about the cedar pole business along the Skeena, which recorded a "remarkable increase" in production to a record 4.5 million feet. Strong American demand prompted Olaf Hanson to extend his network of pole camps east to New Hazelton, and to establish another assembly yard at Nash, east of Cedarvale. The 1929 pole drives carried the up-river output to the booms at Nash, those cut below that point going to the Cedarvale facility further down the Skeena. But the pole industry was not immune from the impending economic catastrophe. Orders fell off immediately in the wake of the stock market crash that autumn, prompting Hanson to lay off his crews and discontinue operations. Perhaps the only solace in forest industry circles in the coming years would be that the Great Depression's impact would be felt less sharply in a region already so familiar with hard times.[71]

Duplex truck towing logs on the Queen Charlotte Islands in the early 1930s.
Vancouver Public Library VPL 16609

4 From Slump to Boom

Depression, War and Changing Patterns in Forest Exploitation

From 1930 to 1945 the northwest timber industry underwent changes that mirrored provincial trends, with a slight difference, perhaps. Because the regional forest economy already suffered from the inability to penetrate domestic and foreign markets consistenly, with the exception of the Ocean Falls paper plant, the northwest had somewhat less to lose than more heavily industrialized areas when the Great Depression hit. All sectors of the industry faltered, but many inhabitants had experience in coping with seasonal and cyclical economic swings by alternating homesteading with fishing, trapping and forestry. The same could be said for resource workers throughout a provincial economy subject to periodic booms and busts, but traditional flexibility may have given those in the northwest particular resilience.

The employment picture was not entirely gloomy, although the pressure to produce increased. The Ocean Falls and Powell River newsprint mills weathered the early 1930s without the calamitous impact felt by lumber interests, and their increasing reliance on fibre from the Queen Charlotte Islands produced heavier logging activity there. Indeed, by the end of the decade, tractors, trucks and even railroads had come into use to reach timber inaccessible from waterfront A-frames. Employers pushed hard to extract every bit of value from workers and the forest, even after the Depression eased toward the end of 1934. Wage cuts, incentive payment plans and efficiency drives in the woods and mills drove workers to their limits. Pacific Mills tightened its grip on contractors and handloggers even further, leaving them with imaginary profit margins. Still, work of any sort was preferable to the dole if one could meet the demand for speed on the job.

The region's sawmills relinquished any hope of participating in the North American rail trade or developing offshore markets. H.R. MacMillan gave up entirely, closing his Big Bay plant and depriving Prince Rupert of sawmill employment. The rest engaged in cut-throat competition for the dwindling number of local orders. Along the Skeena River a barter economy sprang up, with settlers taking lumber in exchange for logs or even produce grown on homesteads. One might also cut a few ties for the CNR or a local contractor, although that option continued to decline along with demand and prices. Recovery of the pole industry by mid-decade offered still another way for aboriginal and non-aboriginal workers to carve out a living in combination with other activities.

Northwesterners managed to survive the Great Depression, then, by relying on time-tested methods of adaptation. The forest weathered the storm too, although more aggressive clearcutting on the Queen Charlotte Islands caused some concern among foresters about what the future might provide in the way of natural regeneration. Good handlogging chances were also becoming harder to locate, and in the interior the increasing distance separating profitable tie and pole timber from the CNR line signalled the depletion of accessible stands. But no one in official circles worried about resource exhaustion; the more pressing problem seemed to be the existence of an abundant resource with no market to consume it.

The arrival of World War II solved that problem, setting the stage for the postwar boom that would draw more fully on up-coast timber. In the short-term, the war triggered another spruce drive and exposed the flimsy scientific foundation of regional forest management. When private operators and the federal government's Crown corporation, Aero Timber Products, turned to selective logging of Sitka Spruce, foresters still lacked real insight into its ecological character. Was selective cutting superior to clearcutting, or merely a form of high-grading the best the Queen Charlotte Islands had to offer? Some 30 years after passage of the Forest Act purportedly placed British Columbia forestry on a sound administrative footing, professional foresters had no answers. Nor had any meaningful strides been made in enforcing sustainable standards of forest practice. This war, like the first, would offer little opportunity for reform. Production, once again, trumped conservation.

The Great Depression had an immediate, although muted impact on the northwest's already depressed forest industry. Provincially the cut fell by 20 per cent in 1930 as domestic consumption faltered and American tariffs began to restrict access to that market. Britain turned to the Soviet Union for lumber, further damaging export opportunities in British Columbia. The total value of production in the Prince Rupert Forest District actually increased 5 per cent to $2,650,594 in 1930, although the industry offered less employment than in previous years. Pacific Mills cut their operation from six days a week to five, continuing to draw most of its raw material from Pulp District No. 1 and the Swanson Bay area. Small operators around Prince Rupert found a market for sawlogs at the Big Bay Lumber Company and Brown's mill on the Ecstall River. Except for an annual maintenance shutdown, the Big Bay plant managed to operate continuously, cutting for the local box-shook market and shipping the better grades to

Sleigh-load of ties, 1930s.
BC Archives B-02150

England and the eastern states. Conditions were not as favourable around Terrace, where the woods industries provided a livelihood for most residents. Cancelled orders prompted George Little, whose Terrace mill remained the most important of the CNR operations, to predict heavy unemployment. C. Pohle and C. Giggey cut intermittently on Kitsumgallum Road nearby, the latter rebuilding at Terrace Siding after losing his uninsured mill to fire. Other plants at Shames and Vanarsdol cut a few orders, and District Forester A.E. Parlow considered it fortunate that the interior mills avoided a complete shutdown in 1930.[1]

Although interior settlers secured most of the tie orders, the "roaming tie hack" finding little work in the winter of 1930-31, district output fell another 20 per cent to 844,030. The Hanson Tie and Lumber Company, which in the past routinely handled contracts for a quarter of a million ties, received an order for only 10,000. Individual settlers were fortunate to have the opportunity to supply the CNR with 500 ties, half of the 1929 amount and only a quarter of the typical 1928 contract. Prices, which had remained fairly constant for several years, suffered a sharp cut. A No. 1 peeled tie worth 70 cents in 1929 earned only 57 cents in 1930. The CNR also tightened its grade specifications to the point where settlers had difficulty finding sufficient suitable timber on their holdings to meet the requirement for top-grade ties. "A great deal of disappointment was felt by all where the industry is of importance," Parlow noted. But he went on to observe that the CNR could probably have got by without any additional ties from the region, looking upon the 1930 contracts "as a measure of relief and means to retain the population along the railway line in B.C.". But perhaps the most ominous structural feature of the tie industry involved its gradual shift eastward after years of culling stands within hauling distance of the line, a process that made the Francois Lake area the centre of activity by 1930. Hanson's cedar-pole operation also suffered a setback, making these the "worst times that have been experienced since the industry began in this District". Eastern buyers, their yards already congested, instructed him to make no further shipments after January. He kept his camps open until midsummer, then shut down with thousands of poles on hand at his Nash and Cedarvale assembly yards.[2]

Although sawmilling remained dormant on the Queen Charlottes, the islands experienced an early-Depression revival in the logging industry. A.P. Allison had to abandon his operations on the west coast of Moresby Island when the log-towing company cancelled its service. He then acquired some timber on the Masset Haida reserve, contracting to employ band members on the falling and bucking crews. After cleaning that up he moved to Sandspit on Skidegate Inlet, conducting the first large-scale tractor logging on the Charlottes. The T.A. Kelley Logging Company produced 31 million feet at Sewell Inlet and Rockfish Bay, operating two new logging trucks over a three-mile road in addition to A-framing with gas donkeys where shoreline timber remained. J.R. Morgan obtained 10 million feet from a Cumshewa Inlet timber sale, shipping his pulp logs to Ocean Falls and Powell River as did Kelley and Allison.[3]

The British Columbia forest industry continued to spiral into depression in 1931, the value of production falling to less than half of the 1929 total. The Prince Rupert Forest District reflected provincial trends, the value of forest products other than pulp and paper dropping 56 per cent from the 1930 level to just over $1 million. Pacific Mills curtailed output from Ocean Falls by 15 per cent, cut wages by 10 per cent, and lowered rental and other living costs for the town's residents. The slump continued to hit the interior portion of the district harder than the mainland coast, where production stayed within 85 per cent of the average of the three previous years. But, as Parlow observed, "unemployment, its causes, effects, and cures, is the most important topic in all centres."[4]

Perhaps the most important development in coastal logging involved Pacific Mills' shift from the Ocean Falls region to the Queen Charlotte Islands, an area that contributed 50 per cent of the Prince Rupert District's 1931 log output. Parlow speculated that the company, having logged most of the accessible timber in Pulp District No. 1 not held under pulp lease, licence and sale, sought to conserve its remaining mainland-coast holdings. Allison, Kelley and Morgan remained the central figures in Queen Charlottes logging. Morgan cut over 14 million feet from timber sales at Cumshewa Inlet in 1931. Allison logged 12.5 million feet at Skidegate Inlet with his tractors, and Kelley took 10.5 million feet from Selwyn Inlet. The Forest Branch's A.B. Hopkinson toured the Queen Charlottes during this period, an inspection that revealed "a very depressing picture" of the region's forest industry. The two large mills at Port Clements and Queen Charlotte City sat in ruins, the machinery intact but with huge piles of lumber "rotting away". The Buckley Bay mill on Masset Inlet presented a less desolate scene, but it too seemed unlikely to operate in the immediate future.[5]

Across Hecate Strait at Prince Rupert the lumber industry exhibited a similarly dismal outlook. The completed Billmor mill sat idle, along with the adjacent National Airplane Spruce plant. Perhaps seeking cheaper logs, the Big Bay Lumber Company no longer obtained its supply under contract from the Queen Charlottes. The firm instead let a number of small contracts in the Skeena River and Douglas Channel area to former Pacific Mills handloggers. But the new arrangement provided only a partial solution to faltering demand, and the mill cut a meagre 5 million feet during several months of limited operation before shutting down that autumn. Desperate to promote any sort of employment, the Prince Rupert Chamber of Commerce repeatedly asked the provincial government to permit the export of hemlock and cedar logs from the region. Realization of the city's dreams of a pulp mill promised a more permanent remedy for the "mass unemployment" that plagued Prince Rupert, and civic officials were tantalized by at least three promotions during the 1930s. A syndicate of North Dakota and Minnesota capitalists were first in line, incorporating the Prince Rupert Lumber, Pulp and Paper Mill late in 1930. They took over the Kleanza mill at Usk and promised a pulp enterprise "some time in the immediate future". F.L. Buckley, "publicity agent" for the group, reported in 1932 that they had come to a "definite understanding" on the establishment of a sulphite mill at Seal Cove. But Parlow had his doubts about the project, which never got off the ground.[6]

The Depression continued to punish the interior forest industry, the only bright spot provided by Department of Public Works orders for the construction of relief camps. Mill owners had no difficulty obtaining inexpensive logs, but bitter competition in the lumber market coupled with the costly rail haul conspired to make profitable operation impossible. "All the mills in the vicinity of Terrace have been shut down for a large portion of the time, opening only for short periods to fill orders," Parlow reported. CNR tie orders fell again to 340,000, at prices attractive "only to the settler who has spare time in the winter when he can make ties instead of being idle". Pole production "shrank practically to the point of extinction". Until the relief camps opened unemployed men holed up in abandoned tie and pole camps, leaving the hotels in Hazelton, Smithers and Burns Lake virtually empty. Considering the unemployment situation in his entire forest district, Parlow thought the interior worse off than the coast, with the Queen Charlottes best able to weather the Depression. The Ocean Falls and Powell River mills kept operating, high-quality spruce still found a market, and while most of the loggers came from Vancouver, the three major Queen Charlotte operators made extensive use of local labour.[7]

The grinding cycle of wage cuts and layoffs continued in 1932, the value of forest products from the Prince Rupert District falling to under

Hand loggers' camp, 1940s.
BC Archives NA-07074

$1 million. Pacific Mills accepted delivery of only 49 million feet of logs instead of the 75 million it demanded in ordinary times. The company slashed wages in the mill by 20 per cent, and spread its contracts around in an apparent effort to keep as many logging operators as possible solvent. The bulk of its mainland log supply, some 30 million feet, came from 29 operators employing steam or gas donkeys around Ocean Falls and Swanson Bay. These men had little option but to compete with one another for the available reduced contracts. "The Pulp Company now has the power to make or break these operators at will, and have taken advantage of the opportunity to dictate terms," A.E. Parlow observed. "Most of the operators did very well to break even." Denied contracts, four steam loggers were forced out of business. Another 8 million feet originated from the labour-intensive efforts of 64 handloggers. Although the Forest Branch reduced stumpage rates on hemlock and balsam, Pacific Mills sought to avoid such charges altogether the following year by encouraging aboriginal and white loggers to take out handlogging licences. Fees and "grub stakes" advanced to these men could be recovered in future, but A.E. Parlow noted the increasing difficulty of locating good handlogging chances.[8]

The coastal sawmills provided little in the way of an alternative market for power contractors and handloggers, experiencing "another bad year" in 1932. J.R. Morgan finally put his Billmor mill into operation in a vain attempt to compete with Vancouver operators in the lumber trade, running only a few weeks before closing without a profit. Even the local markets failed to cooperate; construction stagnated and the Terrace mills cut prices below the cost of production in an attempt to take a share of the fish-box business. Together, the Big Bay Lumber Company, James Brown's Ecstall River mill, the Saloomt mill at Hagensborg, and an Oona River plant operated by J. Hadlund purchased only about 1.9 million feet of logs. The curtailment of logging and milling threw many out of work complete-

Central coast
handlogging, 1940.
BC Archives NA-07075

ly, others eking out a living on part-time wages. A serious cut in prices offered by the canneries and fish-packing plants made this difficult for those who alternated logging with salmon fishing. Only the best-equipped and most skilful fishers managed to do more than cover expenses in the season of 1932, and First Nations women had less opportunity for cannery work as that industry began a contraction that ended the existence of many up-coast plants.[9]

Conditions remained relatively stable on the Queen Charlotte Islands, log production falling by five per cent from the 1931 level. Kelley rafted his entire 21 million feet to Powell River, sorting out the high-grade spruce and selling the pulp logs to the Powell River Company. Allison and Morgan each cut about 10 million feet for the pulp plants and Vancouver mills that purchased prime spruce, Morgan exporting one raft to Puget Sound under permit. Despite wage cuts, Parlow continued to feel that their practice of employing islanders as loggers maintained near-normal employment levels.[10]

Interior sawmill operators adopted desperate strategies to survive as the Depression deepened in 1932. At Terrace, the Little, Van Ardsol and Inter-Valley companies competed hard to fill local demand and enter the coastal box-shook market. "All three mills are working on little, if any margin," Parlow reported, "taking chances on credit, using a barter system with the local people, and generally trying every known artifice to keep solvent." Throughout the interior, barter replaced the cash economy, settlers exchanging logs or agricultural produce for lumber. CNR tie orders fell again, along with prices, to a mere 106,000. A settler might make wages filling a small contract, but Parlow expected that a further price reduction would prompt some to withdraw completely and rely entirely on relief. The few pole shipments drew on existing stocks; only the loggers who hoped to sell their output in the future bothered to cut poles.[11]

Those who had formerly devoted their energies to woods work now had plenty of time for homestead improvement, scrambling for cash by cutting a few ties, raising produce for consumption and sale, or working on one of the province's relief projects. The new devotion to homestead development gave the region "a fairly prosperous appearance", Parlow observed, and despite a scarcity of cash, he now considered the interior's permanent residents to "have entrenched themselves in a better position to withstand the Depression than is found on the mainland coast".[12]

Not until the latter part of 1933 did British Columbia's forest industry begin to show signs of recovery. The large export mills on the Lower Mainland and Vancouver Island would begin to reap the benefits of the 1932 Imperial Preferential Agreement, giving Canadian lumber duty free entry into Commonwealth countries, but this had no immediate impact along the north coast. Still, Parlow thought he detected a "slight upturn in business conditions" toward the end of another dismal year. Pacific Mills operated at under two-thirds of capacity in 1933, making a return on sales in overseas markets but losing money on the 70 per cent of its output shipped to the United States. Still, the operation managed to make a profit even at the bottom of the Depression, earning $36,242 in 1932-33, and posting more impressive results as the decade went on.[13]

The firm exerted its usual dominance over mainland coast logging, contracting for practically all of the area's log production. Handloggers provided much of the cut, as Pacific Mills went ahead with its plan to encourage production from this sector by financing 60 handlogging licences. Sixteen contractors also contributed, although the average contract did not permit efficient utilization of steam or internal combustion logging equipment. Retrenchment at the Ocean Falls mill produced further layoffs and another wage reduction. By this time workers were labouring under a management system introduced by efficiency expert Charles Bedaux, whose services Pacific Mills secured in 1930. His engineers conducted time studies of the employees as the basis for a bonus plan that tied earnings to production. Parlow accurately described this more sophisticated form of exploitation as "a forced-pressure system in which only the younger and fit can maintain the requisite pace to hold their jobs."[14]

The north-coast sawmills, struggling under structural burdens that the Depression only worsened, made no headway in 1933. Shut out of the export business by low prices and intense competition, they fared little better on a restricted local market. The Big Bay Lumber Company at Prince Rupert went out of business, along with J. Hadlund at Oona River. The Billmore plant ran sporadically with a small crew. But the district gained one new operator, and Georgetown came back into production. The

Georgetown Lumber and Box Company began cutting at that site to supply fish boxes and construction lumber for Prince Rupert. K. Kobayashi at Moses Lake, the Salloomt Lumber Company at Hagensborg, Brown's Ecstall River mill and a portable plant operated by Robertson and Simpson at Masset also served nearby settlements and canneries, cutting a meagre total of 213,000 feet.[15]

Having temporarily lost Pacific Mills as a customer, the three major Queen Charlotte logging operators concentrated on rafting premium spruce logs for the Vancouver market and supplying pulp wood to Powell River. Kelley, now logging timber sales and licences at Atli Inlet on Lyell Island, rafted 14 million feet to Powell River. Morgan and Allison recorded smaller outputs from Crescent Inlet and Skidegate Inlet, respectively, but it was Kelley's innovative approach to financing his new Atli Inlet operation that drew Parlow's attention. Rather than pay straight wages, Kelley hired men on the understanding that he would provide board and lodging for the logging season. Wages would be deferred until the logs were sold, involving a nominal rate plus a percentage of any profit.[16]

This variation of bonus payment plans – long favoured by the industry to spur productivity during lean times – typically increased the hazards associated with an already dangerous occupation. "The system apparently was a success insofar as the company was concerned," Parlow reported, "as the men worked long hours and at high speed in an effort to step up production." But the forester expressed more concern for the fate of the resource than the loggers. The Queen Charlotte operators had a well-established reluctance to handle hemlock and cedar except in small amounts, hoping to offset the costs of building Davis rafts and shipping long distances by concentrating on the valuable spruce. Timber of "doubtful merchantability" was left standing, with no care taken to avoid damage during yarding. Kelley carried this policy to the extreme at Atli Inlet, "creaming the two sales ... for spruce only, leaving hemlock, cedar and a percentage of less accessible spruce for possible future operations when prices or a lack of more accessible timber make the undertaking a more attractive proposition."[17]

George Little remained the interior's most prominent lumber manufacturer west of Hazelton in 1933, cutting over 650,000 feet at his main Terrace mill and operating a much smaller facility with C. Pohle. The Vanarsdol, Inter-Valley and Shames Lumber Companies stayed active in that area, and to the east Telkwa, Francois Lake and Fraser Lake boasted mills. Together the 12 interior facilities cut almost 3.5 million feet, Fraser Lake Sawmills topping the list at over a million. Mill hands earned a bare living wage or less, and one could do no better tie hacking. By this time the tie industry was back in the hands of contractors after the CNR's early-Depression experiment dealing direct with settlers. The logging operators, mill owners, and store keepers who functioned as contractors handled orders and sublet 300 to 500 tie lots to natives and settlers who might make

Hansen and Dahlberg pole operation near Terrace, 1934.
BC Archives NA-05397

$3.00 or $4.00 a day for their labour. "About the only parties making a profit out of ties are the head contractors who collect their percentage regardless of the relationship between the prices and logging costs," Parlow commented.[18]

This state of affairs prompted the settlers to attempt to circumvent the contractors by organizing cooperative societies capable of handling contracts directly from the railway. That would eliminate the contractors' overhead and commission charge, which reduced the settlers' profit margin to the vanishing point. Very few poles left the district, Hanson filling orders from yard stocks and hauling those left in the woods over the past two years to the Skeena for driving to Nash and Cedarvale. The Baxter Company withdrew entirely from Masset Inlet, leaving about 1.5 million feet of poles in the yards there to await a stronger market. Like the other sectors, the moribund pole business offered those settlers and natives who participated throughout the region less than a living wage.[19]

The northwest forest industry finally assumed a more healthy posture in 1934, all sectors with the exception of poles showing a marked increase in production. Log output in the Prince Rupert District climbed from the Depression-era low of 56 million feet in 1933 to 136 million feet. Increased buying by Pacific Mills, which operated steadily under stronger paper markets, almost doubled the pace of logging on the mainland coast between Rivers Inlet and Prince Rupert. Mechanized loggers produced nearly 60 per cent of the 58 million feet scaled in the area as Pacific Mills apparently dropped its policy of sponsoring handloggers. Past selective logging of spruce along the coast worked a hardship on some who went handlogging for a few months on either side of the fishing season, the preponderance of low-value hemlock and balsam limiting the gains from many chances.

These conditions encouraged some handloggers to target the Swanson Bay pulp leases still held by the B.C. Pulp and Paper Company. Parlow admitted that negligible amounts had been taken illegally, and in 1934 the firm began patrolling its idle limits to protect them from trespass. Despite policies that allowed B.C. Pulp and Paper to retain control over thousands of acres of timber with no prospect of immediate use, and the capacity of Pacific Mills to suppress log prices, the more buoyant log market provided some short-term benefits for coastal residents whose subsistence depended upon seasonal logging and fishing.[20]

The coastal sawmills remained confined to servicing the local cannery and construction markets, no thought being given to engaging in the rail trade. Skeena River canneries purchased box material from the Georgetown, Ecstall River and Billmor mills. The Salloomt operation at Hagensborg cut for local consumption in the Bella Coola River valley, and the Kobayashi mill supplied the Rivers Inlet canneries. Robertson and Simpson filled the requirements of settlers at Masset Inlet and cut some shingles for Prince Rupert. At Surf Inlet the Princess Royal Gold Mines operated a small mill for construction purposes. Collectively, their log purchases provided no counterweight to the power Pacific Mills continued to exert over prices along the coast.[21]

That firm's revived interest in the Queen Charlotte Islands as a source of fibre contributed to a doubling of the output there in 1934. Some 20 million feet of logs went to Ocean Falls, exerting a positive influence on hemlock utilization. Although J.R. Morgan continued to employ the traditional "A-frame, skyline and cold-deck" methods of logging on the Charlottes, the exhaustion of shoreline timber along the major inlets set the stage for the introduction of new transportation technologies. Allison followed the usual practice at Cumshewa Inlet, operating three cold deck settings to reach the back of his limits. But he also initiated construction of a railroad from the Inlet to Skidegate Lake, through Powell River Company timber anticipated to afford at least 10 years of logging. T.A. Kelley had already begun using trucks to extend his reach, in addition to logging waterfront timber with A-frames at several points.[22]

Although interior sawmillers still found themselves shut out of the prairie lumber market, and the region's pole industry showed no improvement, 1934 CNR tie purchases increased to over 400,000. Moreover, the firm's growing preference for sawn ties over the hewn variety, in conjunction with the ever-growing distance separating suitable stands from the line, provided new opportunities for mill owners. Portable tie mills powered by internal-combustion engines also appeared in increasing numbers in the bush, processing Western Redcedar, Western Hemlock and Jack Pine timber. Mill-sawn ties constituted almost 25 per cent of the 1934 output, the bulk still taken out by settlers hewing ties under sub-contract to Hanson and other contractors. "Providing a settler has made some headway on the land, any means of acquiring say $500.00 in ready cash will bring him prosperity," Parlow thought. "Hence, a tie contract for a thousand ties is nearly sufficient for his needs providing he makes and hauls the ties himself."[23]

The mid-point of the Depression brought no fundamental change to the structure of the northwest forest industry, a slight decline in log production being offset by an upward price trend. Logging between Prince Rupert and Rivers Inlet fell off as Pacific Mills increased its purchases from the Queen Charlottes and Seymour Inlet, just across from the northern tip of Vancouver Island. That development boosted the rate of cut on the Charlottes to 82 million feet, up over 10 million feet from the 1934 output. The introduction of a new bleaching process for hemlock at Powell River limited that organization's purchases of pulp spruce and hemlock, forcing the operators there to maximize their sales to Pacific Mills. A.P. Allison temporarily abandoned his plans for a railroad in the Skidegate Lake area in response to the Powell River Company's diminished interest in the cheaper grades, and Parlow expressed concern that the Ocean Falls mill would now be in position to extend its control over the log market from the mainland coast to the Charlottes.[24]

Parlow suspected that the Queen Charlotte operators would apply for permission to export their low-grade spruce and hemlock, a prediction that proved accurate in 1936 when J.R. Morgan sold three rafts of spruce to a Tacoma mill and shipped four million feet of spruce and hemlock to Japan. For Parlow, a Prince Rupert pulp mill offered the only remedy for a host of regional economic ills. "A pulp plant at Rupert would not only greatly benefit the town directly," he pointed out, "but would make a local log market, and by creating competitive buying, would improve the situation for all the loggers on the Northern Coast and Queen Charlotte Islands."[25]

And for a time during the mid 1930s the outlook for such a transformation seemed bright. F.L. Buckley had not given up, serving as managing director for the Canadian-American Pulp and Paper Company, incorporated in 1935 to build a sulphite plant at Prince Rupert. The promoters held an estimated four billion feet of timber on the Charlottes, and the Skeena and Nass rivers, but 1936 passed with no construction. Other groups investigated the potential for northwest development. A party of Germans visited in the spring of 1936, another representing Eastern Canadian and British capital made an inspection tour later that year, and an American company expressed an interest in Prince Rupert. Parlow confessed at the end of 1937, "the much-mooted proposed pulp mill for Prince Rupert is far from being an accomplished fact," although one syndicate had made further visits to study raw material supplies, power prospects and possible mill sites. A hopeful Prince Rupert City Council had even passed attractive taxation by-laws.[26]

Parlow remained convinced that real growth in the region's logging sector would only come through the establishment of another pulp mill. Pacific Mills' capacity to dictate prices prevented operators from developing anything other than low-cost handlogging and A-frame shows. A competitive log market would promote a rise in log prices, allow loggers to open up larger operations and "greatly improve conditions for North Coast operators". But pulp-and-paper industry profits plummeted in the wake of the "Roosevelt recession" in 1938, a downturn that removed much of the incentive for investment in another northern British Columbia mill. "Possibly this accounts for the fact that during the year the customary unkept promises for the early erection of the mythical Prince Rupert plant were not reiterated," a forester remarked.[27]

In the absence of a new pulp mill coastal loggers remained at the mercy of Pacific Mills, which absorbed 41 million of the 43.3 million feet of logs produced along the northwest coast in 1936. The opening of company camps on South Bentinck Arm that spring further undermined opportunities for handloggers and contractors, providing almost 12 million feet in 1936 and another 15 million feet in 1937. When the Depression returned with a fury in 1938 the firm closed its camps – this would have been a fortunate break for the small coastal operators, but it turned sour when Pacific Mills reduced its log consumption that year from 74 million to 50 million feet. The Ocean Falls mill curtailed production and shut down entirely at various times in an effort to cope with fiscal conditions "on a par with the depression years of 1933–34".[28]

Not until the summer of 1939 did Pacific Mills resume normal production levels, but the reopening of company camps at South Bentinck Arm and Elcho Harbour that year again limited its demand for the output of the independent loggers. Low log prices and diminishing timber suitable for A-framing continued to wear away at the contractors. Fewer and fewer could afford the equipment necessary to conduct lengthy cold-decking and swinging operations to bring logs to tidewater. At the same time Pacific Mills discontinued its policy of financing aboriginal handloggers. "Too many did not deliver sufficient logs to cover advances made," a Forest Branch representative explained, but he predicted that wartime economic revival would spark renewed activity among handloggers.[29]

The pace of forest exploitation picked up in the Bella Coola River valley during this period in a process marked by debt accumulation and business failure. Having acquired cutting rights on numerous Crown grants that offered export privileges, the Bella Coola Timber Company began selective logging Douglas-fir timber with "cats" and trucking logs over a 12-mile road to tidewater in 1937. The company planned to log 12 million feet per month for export by freighter to Japan, but economic circumstances and mismanagement doomed the enterprise to failure. Log output fell behind schedule, and when the Japanese market vanished the firm rafted the logs to Vancouver. When their arrival coincided with a drop in fir prices, the enterprise folded.

By 1938 the Viking Timber Company of Mission had taken over the assets with an ambitious plan that included construction of a large sawmill at Bella Coola. In 1939 the firm took out the timber left by the previous operator, put another million feet of fir in the water, shipped one barge of spruce to Washington state, and did some sawing in a small portable mill before it "folded up engulfed in debts and wage liens". By midsummer a group of local residents had organized a cooperative to assume the Viking Company's assets and liabilities. The Northern Cooperative Timber Association took advantage of a rising market, logging steadily through 1939.[30]

Despite a diminishing number of virgin tie stands, most within six miles of the railway having been selectively logged by 1935, labour conditions improved slightly along the Skeena River during the mid 1930s. Settlers and natives delivered 453,000 hewn ties to the CNR in 1935, and a number of portable mills cut another 135,000. Railway orders dipped slightly in 1936 and fell again in 1937 to 253,000, of which 36 per cent were sawn by portable mills. These homemade affairs proved essential in utilizing larger, rougher timber left by the big pre-Depression tie camps. Since neither prices nor the magnitude of orders justified building roads into stands more distant from the railway, the most successful operators stationed their small mills on previously cutover areas to saw on a piece-rate basis logs hauled and decked by settlers. Foresters considered this more intensive pattern of utilization worth encouraging. Although the scattered timber and long skidding distances ate into earnings, the mills eliminated backbreaking broadaxe work and made use of trees rejected during previous hewing operations.[31]

Even the pole industry showed some recovery in 1935, shipments of over one million feet absorbing existing yard stock and prompting new operations between Terrace and Moricetown. Production increased slightly over the next two years, and this sector too saw operators comb through cutover areas rather than go to the expense of building new roads and camps. Aboriginal labour figured prominently in pole production around Hazelton, whites predominated at Terrace, but both groups saw little return for their labour. Only the new demand for sawn ties gave interior sawmillers relief from the Depression. Local building projects generated some orders for the roughly 30 small mills east of the Cascades in the Prince Rupert Forest District, but few export opportunities existed. Industry veteran George Little sold his Terrace lumbering interests to Duddley Little, C. Haugland and Duncan Kerr late in 1935. Terrace was the only remaining locality that offered timber suitable for manufacture into products attractive to outside buyers. "The only stands which might be classed as lumber timber centre around Terrace," remarked a forester, "and while of fair quality, cannot in view of the additional freight rates compete in the Prairie or Eastern markets with the Spruce from the Prince George country or the cedar from the lower coast."[32]

The return of healthier pulp-and-paper markets gave rise to aggressive logging on the Queen Charlotte Islands at mid decade. The "Big Three" employed about 400 workers on the Charlottes during this period, shutting

A.P. Allison Company log dump at Cumshewa nlet, 1938.
BC Archives NA-10958

down each December for the Christmas break and resuming operations
when weather permitted in the new year. Production hit 93 million feet in
1936 and a new record of 105 million the following year before falling back
to 75 million in 1938 as Ocean Falls and Powell River cut back their log
consumption. Tom Kelley's utilization practices on Lyell, Tanu and
Richardson islands came in for praise from Parlow. Directing his Western
Hemlock to Ocean Falls and Sitka Spruce to Powell River, he cut the high-
grade portion of the latter species in the sawmill he leased from the Powell
River company, sending the poorer logs to the paper plant. Allison revived
his Cumshewa Inlet railway project through Powell River Company timber, putting in two miles of line by the end of 1936 and hauling logs over
four miles of track to tidewater in 1937. Equipment purchased from Pacific
Mills helped Allison increase his output from 18 to 31 million feet during
this period, still less than the 41 million feet logged by J.R. Morgan at
Thurston Harbour in 1937.[33]

Morgan raised eyebrows by adopting "cat" logging on the Queen
Charlottes that year, an experiment that proved successful despite muddy
conditions produced by heavy rainfall. But Allison's connection to the
Powell River Company timber set his operation apart from those run by
Kelley and Morgan, who continued to move frequently in search of cutting
rights. Morgan demonstrated his flexibility in the 1938-39 period, first
moving his operation from Thurston Harbour to Pacofi in the autumn of
1938. By the next spring he had 22 million feet felled and partially cold-
decked there, much of it hemlock marketable only at unattractive prices.
He abandoned those logs for the time being, moving back to Cumshewa
Inlet where he had acquired a block of licences showing a larger quantity
of spruce. He logged there until the hemlock market picked up later in
1939, then shifted most of his equipment back to Pacofi. Looking ahead to

J.R. Morgan's tractor logging operation, 1938.
BC Archives NA-10968

at least 10 years of logging at Cumshewa Inlet, Allison was able to invest in a permanent camp at Deland Bay for his 160 loggers.[34]

By 1937 just over 19,000 acres had been logged on South Moresby and the adjacent islands, over half during the previous five years. Although the concentration on high-grade spruce close to tidewater had produced a pattern of scattered small clearcuts with sufficient residual timber to permit natural restocking in most cases, the recent trend toward large-scale logging had left "more extensive cutover lands with a less satisfactory seed supply". Moreover, shoreline logging shows were becoming increasingly scarce, evident in the adoption of trucks, crawler tractors and even a railroad to reach inland timber.[35]

A 1937 Forest Branch survey of Moresby Island concluded that the rate of cut had reached the approximate "allowable sustained yield capacity". Satisfactory regeneration was apparent on 80 per cent of the area logged and burned in the past 20 years, but only 50 per cent of the cutovers not exposed to fire had achieved an acceptable level of restocking. Dense slash, Salal, berry bushes and unsatisfactory seed supply appeared to be the main factors inhibiting forest renewal, but the entire question of cutting policy and silvicultural treatment required investigation. In the interim, the survey's author recommended that future timber sale contracts impose a patch logging system that would limit clearcuts to 40 or 50 acres and require the leaving of scattered seed trees to promote natural regeneration. Selective logging with tractors should also be encouraged where appropriate, provided that the practice not be "abused for the sole purpose of high grading".[36]

By the mid 1930s, the provincial government's proper regulatory role in relation to the forest industry had become a subject of intense public debate in the southwestern corner of the province, where decades of unrestrained clearcutting had left residents of timber-dependent communities worried about their future. Massive clearcuts denuded entire valley bottoms on Vancouver Island and the lower coast, leaving a fringe of high-elevation timber to re-seed cutovers. Over 600,000 acres of cutover land in the Vancouver Forest District stood barren, and another 400,000 acres exhibited minimal restocking. Only 25 per cent of the cutover private land in Vancouver Island's Esquimalt and Nanaimo Railway belt featured satis-factory regeneration. On private land and temporary tenures alike opera-tors sought "rapid liquidation of their timber assets", a forester concluded in a plea for reforms.[37]

Taking advantage of a wave of public concern over the devastation left by industry's mass production regime of railroads and overhead logging systems, E.C. Manning proposed a moderate reform agenda after becom-ing Chief Forester in 1935. His campaign produced 1937 amendments to the Forest Act giving his agency the power to compel slash burning on Crown grants and temporary tenures, and to proceed with seed tree regu-lations as an agent of reforestation. The results, in the end, would fall far short of the hopes of those who advocated strenuous state intervention in the affairs of private enterprise. For the most part what Jeremy Wilson calls industry's "liquidation project" went on unimpeded by governments more dedicated to promotion than regulation.[38]

Further research is required to determine the extent to which residents of the northwest shared the conservationist concern evident in the more industrialized centres to the south, but among foresters little is evident dur-ing the Depression. Indeed, the region's resources seemed to be wasting away, awaiting the sort of industrial development needed to generate new settlement, jobs, profits and revenue for the Crown. A summary of the 1931 inventory of the Prince Rupert Forest District noted that "utilization has been comparatively little so far", merchantable timber occupying 54 per cent of the productive forest area. Another 15 per cent held fully stocked stands of natural reproduction, with sparse stocking on 11 per cent of the area. Fires east of the Coast Mountains had left 20 per cent of the produc-tive land without a new crop, but the remaining volume of saw-timber stood at an impressive 58,870,000 feet.[39]

This total represented a reduction of only 8,410,000 feet from the estimate prepared by the 1917 Dominion Commission of Conservation. Not all of this timber was accessible under current operational practices, but according to the survey the average annual total depreciation by log-ging and fire amounted to only a third of what could be cut on a sustained yield basis. Sitka Spruce and Western Hemlock comprised 55 per cent of the total volume, a wealth of pulp species "expected to have considerable influence on the nature of future industries using timber from this region". Overall, then, the Forest Branch considered the Prince Rupert District to hold a vast surplus of timber, one likely to gain importance given the over-cutting of Douglas-fir forests in the Vancouver District.[40]

F.D. Mulholland's 1937 report, *The Forest Resources of British Columbia*, presented a similarly rosy view of the northwest's resource status. Together the coast and interior forests were being depleted at a rate of 155 million feet a year, "little more than half their capacity". Growth of the interior forest exceeded depletion by about 250 million feet annually, but most of this area had yet to become accessible. Even on the coast, scene of the most intensive logging, the accessible forest could yield another 100 million feet annually. The abundance of pulp timber would easily support a "considerable increase" in the district's pulp-and-paper industry beyond the existing Ocean Falls mill. But Mulholland also noted that south coast operators had begun harvesting the Prince Rupert District's timber, creating a "drain" to Vancouver-area mills that would need to be checked if the region's industrial potential was to be realized.[41]

Although the northwest's timber abundance bred complacency among Depression-era foresters, the pace of forest exploitation was about to pick up dramatically. The outbreak of World War II in September 1939 ended the lingering Depression, bringing boom conditions to the provincial forest industry. An initial shortage of shipping and German submarine raids limited access to the British market, but the war effort soon created an enormous demand for wood products. After negotiating a large order from Britain's Timber Control Board, H.R. MacMillan accepted the position of timber controller under Minister of Munitions and Supply C.D. Howe on the War Industries Control Board. MacMillan, biographer Ken Drushka asserts, quickly "transformed the fiercely competitive BC lumber industry into a cooperative war effort", organizing production to meet British and Canadian government needs.[42]

Conditions began showing improvement along the north coast in August 1939, and by the time war erupted Ocean Falls had all five of its paper machines running. With Britain cut off from its Baltic timber supplies, the H.R. MacMillan Export Company looked into the possibility of exporting hemlock lumber to the United Kingdom through its Canadian Transport Company subsidiary. But the lack of an adequate lumber assembly wharf and loading facilities at Prince Rupert made profits unlikely, killing MacMillan's enthusiasm. With this opportunity gone, the Georgetown, Brown and Billmor mills continued to produce small amounts of lumber for local construction and box shooks for the fish plants. Rumours circulated at the end of 1939 that Morgan's idle Billmor mill would never operate again.[43]

Tie and pole production continued to outweigh sawmilling in the interior as the district prepared for the wartime economy. Another dip in tie orders worried settlers around Smithers, Burns Lake and Southbank, as the CNR confined itself to meeting maintenance needs. Although 1939 pole shipments increased to 1.8 million feet due to orders from the railway and American buyers, these drew mainly on old stocks. By this time, about 70 per cent of pole production originated in the Hazelton district, the Terrace

Booming grounds at Ocean Falls, 1940.
BC Archives NA-07717

area supplying the rest, but prices would have to rise to make their reserves of pole timber economically accessible. The Forest Branch estimated that in 1939 tie and pole cutting generated 33,000 man-days worked for settlers, and another 3,500 for natives, generating $146,000 of income between Terrace and Endako. Although the average $4.00 per day earned on these piecework and contract operations seemed meagre, they continued to supply essential cash for settler families during the winter months.[44]

It wasn't until 1940, "a banner year for logging and manufacture of timber products", that the Prince Rupert District experienced the full effects of the wartime boom. The agency's scale of sawlogs increased to 223 million feet, 80 per cent over the 1939 total, breaking all production records. Pacific Mills ran at full capacity over the entire year, producing 450 tons of paper a day on logs supplied by contractors and a new company camp on Kwatna Inlet. The records fell again in 1941, when the value of production reached $4.2 million. Labour and equipment shortages lowered the 1942 output, but 1943 production hit a new high of 245 million feet, a rate of growth surpassing that for any other area of the province.[45]

Central to the northwest's increased production was another war-inspired assault on the Sitka Spruce of the Queen Charlotte Islands. The area produced 70 per cent of the Prince Rupert District's 1940 log cut as

A-frame operation, 1939.
BC Archives NA-11557

Fore-and-aft road, Queen Charlotte
Islands, 1940.
BC Archives H-05405

the demand for airplane spruce began to climb. Morgan logged at Sewell and Selwyn inlets in addition to maintaining his tractor camp at Cumshewa Inlet. After cleaning up his Tanu and Richardson islands licences Kelley initiated truck operations on Limestone Island and the north shore of Cumshewa Inlet. Kelley resorted to trucks only to access timber beyond the reach of A-frame and cold decking operations, building roads up to three miles long in advance of logging. Trucks equipped with solid rubber tires ran along fore-and-aft roads constructed of hewn timbers resting on logs, a transportation system necessitated by thin soil and heavy rainfall. Allison extended his rail-line eastward on the south side of Skidegate Lake during this period. Pacific Mills also moved into the Skidegate Inlet region, logging high-grade spruce selectively with tractors for its Ocean Falls sawmill.[46]

Rising spruce values prompted the paper company to shift more of its logging to the Charlottes in 1941, issuing fewer contracts to mainland coast operators in order to handle pulp material produced in conjunction with its spruce program. A high-lead operation took the place of the selective logging camp at Skidegate Inlet, the latter shifting to Masset Inlet in midsummer. The Allison Logging Company opened up a rail-line at the west end of Skidegate Lake, but the Morgan and Kelley companies continued to move in search of accessible timber. After exhausting his claims at Pacofi, Sewell and Selwyn inlets, Morgan shifted his camps to Burnaby and Huxley islands in preparation for the 1942 season. Kelley relocated to Sister Creek after wrapping up at Cumshewa Inlet and Limestone Island.[47]

By 1942 the Allied war effort was in high gear, creating unprecedented demand for spruce in the developing aviation field. Pilots from around the Commonwealth arrived in Canada to learn their trade under the

St Faith towing Davis raft for Allison Logging, 1941.
Vancouver Public Library VPL 6270

auspices of the British Commonwealth Air Training Plan. Although metal had replaced wood in many aircraft by this time, trainers still utilized spruce for components such as struts, wing frames and spars. Production of Britain's Mosquito bomber, fabricated almost entirely of wood, made Queen Charlotte Islands spruce essential to Britain's war effort. In order to accelerate production of the vital material the federal government created a Crown corporation, Aero Timber Products, in June 1942. Eventually Aero would direct the operations of all the Queen Charlotte companies, buying Allison's operation outright.[48]

The World War II spruce drive, like that in World War I, would sacrifice conservation to all-out production. But it also coincided with mounting concern within the Prince Rupert District office about the agency's ignorance of the best way to manage its valuable spruce holdings. The annual cut of spruce already exceeded its sustained-yield capacity on Moresby Island by 10 million feet at the war's onset, and the Forest Branch had not yet conducted the sort of silvicultural study required to promote reproduction of the species. Foresters knew that its seedlings were shade tolerant, though less so than either Western Redcedar or Western Hemlock as time passed, but lacked insight into the impact of cutting methods. "Under present methods of logging obtaining successful regeneration of spruce is a doubtful problem," was all the Prince Rupert District could offer on the matter in 1939.[49]

Indeed, the entire relationship of logging practice to forest renewal in the northwest generated a pessimistic outlook on resource sustainability. "I regret that I can see no change over past years in this regard," District Forester R.C. St. Clair reported in 1940. "Our mature stands are still being cut from an economic rather than a silvicultural standpoint, and where the two happen to more or less work towards a common end, such result has been accomplished without any conscious planning and is purely

High-lead logging airplane spruce, Queen Charlotte Islands, 1940.
BC Archives NA-07031

coincidental." Competition in world markets, the traditional explanation for the province's primitive forestry standards, no longer held up given its near monopoly in the British and Commonwealth markets. Now the pressure of war justified inaction, but the Timber Controller's vast power to regulate industry and prices might be extended to achieve progress in utilization and silviculture, St. Clair suggested.[50]

Neither provincial nor federal officials had any interest in subordinating production needs to long-term sustainability during the conflict, especially after E.C. Manning's tragic death in a 1941 airplane crash while serving as MacMillan's Assistant Timber Controller. His successor, C.D. Orchard, would lay the foundation for the province's postwar conversion to sustained-yield management with a 1942 report proposing a cooperative tenure system that evolved into the Tree Farm Licence, but policy innovation would have to await the outcome of the war. On the other hand, some speculated that industry's market-inspired experimentation with selective logging would produce silvicultural benefits. The crawler tractor, capable of extracting high-value timber from mixed stands, provided a technological alternative to the overhead systems which typically mandated clearcutting.[51]

Thus, foresters took a strong interest when Pacific Mills began tree-selection logging high-grade spruce with tractors at Skidegate Inlet. The ranger on the scene in 1940 considered the results disastrous, describing the approach as "a very poor, uneconomical and unsatisfactory method in all phases." Selecting only the best timber for logging, and removing only the clear portions of each tree, the company simply left the remainder to rot. The damp climate reduced the fire hazard on the resulting sea of slash, a major concern on the south coast, but that blessing fell short of convincing the ranger. "When taking into consideration the damage caused to standing timber by broken tops, broken limbs and deep stem scars," he advised, "it is regretted that a further stumpage increase cannot be put into effect."[52]

But St. Clair had no such reservations after a cursory inspection of one or two areas from the agency's launch. The residual stand appeared "in remarkably good condition", with no evidence of a "broken and ravaged understorey". The area could be logged again in 15 or 20 years, an impossibility after clearcutting. Moreover, that procedure produced an equivalent degree of waste, and the relatively shade-tolerant spruce, cedar and hemlock species could be expected to reproduce below the remaining timber. Finally, St. Clair thought that removal of the old-growth spruce would not cause excessive blow-down in residual timber. "It seems to me that tree selection on the Queen Charlotte Islands should be a good silvicultural practice," he concluded. But these were tentative conclusions at best, he acknowledged in recommending that the area receive the same sort of research attention accorded the Douglas-fir forests on the south coast. "We should have the information at hand on which to decide what type of logging to encourage and what to avoid," St. Clair pointed out.[53]

The need for such study became even more evident the following year, when the District Forester confessed that his initial enthusiasm over Pacific Mills' selective logging was unjustified. Severe winds had caused extensive wind-throw in the residual stands, and the company's disregard of low-grade material left an enormous amount of waste. Pacific Mills moved a high-lead operation in to re-log the area, but the district foresters had an even more pressing problem on their hands in 1942 when Aero Timber Products launched large-scale selective logging of high-grade spruce and hemlock at Masset and Skidegate inlets. The Crown corporation would be concentrating solely on aero-grade material, "leaving or destroying all other timber on the areas logged." Even more alarming was the prospect that the other Queen Charlotte Islands operators would have to conform to Aero's practices, "a conversion that will leave many problems for us in its wake."[54]

Aero Timber Products' headlong pursuit of maximum spruce production, conducted with utter disregard for forestry principles, took an even higher toll than anticipated. Adapting high-lead technology to selective logging in its Queen Charlotte operations, the corporation left stands in

Felled timber at Pacific Mills' Sandspit operation, 1945.
BC Archives NA-08432

horrible condition marked by "tremendous wastage of timber in tops, broken and uprooted trees, and trees pulled over in yarding." Most other operators followed Aero's lead, although by the end of the war Kelley, Pacific Mills, and Morgan had adopted a less destructive "group selection" method. The conclusion of the war saw Aero's assets, including trucks, roads, donkeys and railway stock sold off to the established producers on the Charlottes.[55]

While foresters looked with deep concern upon the silvicultural impact of the second spruce drive on the Charlottes, a boom of similar proportions occurred along the Skeena River. Although the CNR's freight-rate structure continued to place northwest lumber producers at a disadvantage to mills east of Prince George, defence projects at Prince Rupert and Terrace generated a burst of construction activity in 1942. The arrival of Canadian and American forces for joint defence of the Pacific coast after Pearl Harbour boosted Prince Rupert's population from 6,500 to 21,000 by 1943, and an estimated 5,000 troops at Terrace demanded the construction of barracks, training facilities and a hospital. Unfortunately, limited access to log supplies restricted operations of the small mills around Prince Rupert. The B.C. Bridge and Dredging Company took over Morgan's Billmor plant temporarily, enlarging the facility to supply the U.S. Army Corps of Engineers.[56]

In the interior defence projects created a "voracious local market", but operators now had trouble attracting labour due to the higher wages paid by contractors building the Terrace-Prince Rupert road. "Old men, Indians, and in a few cases women, are filling jobs formerly carried on by experienced loggers and mill men," St. Clair reported in summarizing the resultant inefficiency. Aboriginal men in particular gained access to woods work in larger numbers, but not, apparently, at the expense of traditional economic and cultural pursuits. "Their continuity of employment is always questionable," the 1942 District report complained in reference to absenteeism for winter ceremonial activities.[57]

By the end of 1943 over 60 small mills operated east of Terrace, filling defence orders along the Skeena early in the year before rail shipments eastward began absorbing most of their output. Over the following year the number of mills in that area almost doubled, bringing the log scale from 23 million to over 36 million feet in response to new opportunities in prairie and eastern markets. A number of trends continued to depress hewn tie production, however. Railway construction had stalled, and while the CNR had to replace perhaps 10 per cent of its ties annually, the use of more and more creosoted ties lowered this number each year. Increased wartime traffic over the CNR provided a need for extensive maintenance and higher tie prices, but the mill-sawn variety grew proportionately greater. Interest in the field waned as well, with few of the old-time "tie-hacks" remaining active. Enlistments and war industries had reduced the number of young men willing to take up a task that, in the best of times, provided thin remuneration for the effort and skill required. "While a poorer class of labour can be used in sawmills, it takes a pretty good man to hew ties and make anything out of it," the District Forester remarked in explaining the gradual passing of the enterprise.[58]

Interior pole production continued its steady upward trend in the early war years, but by 1944 more attractive wage-earning opportunities in conjunction with the imposition of price controls prompted most Euro-Canadians to go elsewhere. First Peoples continued to figure prominently, re-logging areas for timber rejected as too defective or too small in previous operations, but they too would exert their independence. Hanson, Little, Haugland and Kerr, Carl Pohle, and C. Dahlquist topped the list of contractors whose ability to attract aboriginal pole cutters was compromised by rising fish prices as the war went on. And by the end of the war, depletion of suitable cedar pole timber within reasonable hauling distance to CNR sidings around Terrace forced operators to reach into new territory. Little, Haugland and Kerr began opening up a tract north of Kitsumkalum Lake in 1945 even though buyers proved willing to accept poles once considered substandard.[59]

Despite a dramatic shift in economic context, the years between the beginning of the Great Depression and World War II brought no fundamental change in timber capital's relationship to the northwest forest environment. Corporate executives, small operators, handloggers and government officials all shared a utilitarian view of the forest despite debate

Workers dump logs at Little, Haughland and Kerr Mill, Terrace, 1945.
BC Archives NA-08416

about how the wealth generated from its exploitation should be distributed. First Nations shared a quite different view of nature, but a relationship premised on respect and cultural sanctions against over-exploitation had no influence in policy-making circles. And if the two systems came into direct conflict, as they did during the 1930s and 1940s when the Forest Branch suppressed aboriginal burning practices that encouraged growth of berry patches, the aboriginal one gave way to the market-driven conception of nature.[60]

Cultural conflict of this sort passed largely unnoticed except among those whose subsistence practices were suppressed. The challenge thrown up by workers to timber capital's authority to unilaterally determine the conditions of work and life in British Columbia's logging camps and mill towns would not be so easily ignored. The north coast would play a pivotal part in this struggle during World War II, when the urgent need for Sitka Spruce from the Queen Charlotte Islands gave the International Woodworkers of America the leverage needed to gain a foothold in the struggle for collective bargaining rights. Propelled by victory in a 1943 strike in the spruce camps, the union went on to secure its place in the province's political economy by the end of the war.

5 "No Camp Large or Small Will Be Missed"

The IWA and the Loggers' Navy, 1935–45

The story of unionization in the coastal forest industry between the 1930s and the end of World War II is a familiar one to students of British Columbian labour history. This account attempts to sharpen our understanding by considering the conditions of life and work in up-coast camps, the challenges faced by union organizers in spreading the word to these isolated industrial sites, and the pivotal wartime role of the Queen Charlotte Islands Sitka Spruce camps in the achievement of collective bargaining rights. Central to this process was the development of the International Woodworkers of America's (IWA) Loggers' Navy, vessels that enabled the union to extend its organizational reach up the rugged coast to the Queen Charlottes.

After the Imperial Munitions Board's World War I drive for airplane spruce unleashed a brief period of industrialization, the Queen Charlotte Islands became a fibre producing site for the newsprint plants operated by Pacific Mills at Ocean Falls and the Powell River Company. By the 1930s T.A. Kelley, J.R. Morgan and A.P. Allison had become the "Big Three" of Queen Charlotte logging, rafting valuable spruce and less profitable hemlock logs to those points from pulp company leases and timber sales. By 1934, as the British Columbia forest industry began recovering from the worst years of the Depression, log production on the Charlottes reached 82 million feet. Output continued climbing to 105 million feet in 1937 before the "Roosevelt recession" caused paper production to fall off briefly.[1]

World War II brought an immediate revival, thanks to booming pulp-and-paper markets and a demand for aircraft construction material. Sitka Spruce, flexible and light in weight, held an important place in the Allied war effort as a component of many military aircraft, including Britain's Mosquito bombers. The 1942 creation of Aero Timber Products, a federal Crown corporation designed to spur output from the Charlottes, reflected spruce's prominence even after the development of light metals. Aero would operate its own camps, in addition to overseeing production of aircraft material by Kelley, Morgan and Allison, who continued to supply the paper mills throughout the war. Pacific Mills established its own camps as well, and together the spruce camps provided the IWA with an essential organizational focus.[2]

The union faced a daunting challenge elsewhere along the north coast, where Pacific Mills drew logs from numerous small contractors and hand-loggers throughout this period. Small float-mounted A-frame operations shifted from site to site cutting shoreline timber, driven hard by the paper company's ability to set prices. Pacific Mills ruled the north coast by virtue of its timber holdings and the absence of a regional sawmilling sector to generate a competitive log market. Headquartered in Vancouver, the IWA would have to find a means of reaching out to crews in long, isolated up-coast inlets and along the shores of the Queen Charlotte Islands if it was to achieve its goal of drawing the entire coastal industry into the fold.

The difficult job of organizing coastal woodworkers began in 1930, when the Communist Party of Canada created the Workers Unity League (WUL) to promote unionization of mass production workers ignored by the mainstream labour movement. On the west coast, a small group of activists revived the Lumber Workers Industrial Union (LWIU) as a WUL affiliate, obtained organizing materials, and began the daunting task of spreading the gospel of industrial unionism among loggers and mill workers in the dark days of the early Depression. It would be tough going indeed in a time of high unemployment, unfriendly labour law and employer hostility.[3]

Early efforts concentrated on the industrial centres of the Lower Mainland and Vancouver Island. Future IWA leader Harold Pritchett played a prominent role in a 1931 strike at the huge Fraser Mills sawmill that forced concessions from ownership. Loggers on northern Vancouver Island struck the following year, and by early 1934 the LWIU felt ready to take on the logging operators for higher wages, better safety and living conditions, and union recognition. Loggers in the largest island camps stayed out three months that spring, a conflict that produced a wage gain, 3,000 new members, but no acknowledgement of the LWIU from employers.[4]

Over the next year and a half the LWIU and boss loggers skirmished – the latter strengthening a blacklist designed to bar activists from the camps – while the North American labour movement underwent momentous changes. In the United States John L. Lewis led the way in establishing the Congress of Industrial Organizations (CIO) to organize industrial unions that crossed traditional craft lines favoured by the American Federation of Labour (AFL). The CIO would take advantage of Franklin D. Roosevelt's New Deal, its 1935 Wagner Act conferring the right to organize and bargain collectively. Meanwhile, the rise of fascism in Europe prompted the Communist International to disband its independent unions and again work within labour's mainstream. By the spring of 1936 the LWIU was gone, replaced by the Lumber and Sawmill Workers Union of British

Columbia (LSWU) in affiliation with the AFL's United Brotherhood of Carpenters and Joiners (UBCJ).[5]

By this time the union was again locking horns with operators for higher wages and recognition. The dispute, centred at Cowichan Lake, took in north-coast loggers whose grievances included pay rates, food, living conditions, safety and medical care, and the cost of travelling by steamship between Vancouver and the camps Queen Charlotte Island operators, with the apparent exception of Allison, had developed a relatively tolerant attitude to organizers leading up to the 1936 strike. Workers at Morgan's camp could purchase the *B.C. Lumber Worker* from the union's agent, although he was not allowed to sell the paper in the bunkhouses. A delegate had carried out his duties "more or less openly" for three years at Kelley's camp before being discharged that spring. A correspondent attributed the firing to the recent partnership between the Carpenters and the LSWU, a reasonable assumption given that organizers at Cowichan Lake had just been discharged for advocating the merger and selling the *Lumber Worker*. That event triggered a walkout in neighbouring camps, leading the LSWU to call a general strike in May that eventually involved 2,000 coastal loggers and produced an early June settlement for a 50 cent wage increase.[6]

Conditions in the north-coast logging camps, many of which sat on rafts, defy easy generalization. "All the larger camps are electric lighted, washing, bathing, sleeping, accommodations are good, and the meals are excellent," District Forester Parlow reported in 1936. This rosy picture had some validity at Pacific Mills' Carstairs float camp on South Bentinck Arm. Clean bunkhouses, lit by a single gas lamp, held six men each, and plenty of hot water was available in the wash house. Board, described as "fairly good", cost $1.20 a day, plus another 10 cents for blankets. "On the whole, this camp is as good as you will find on the coast," one logger wrote. Other smaller Pacific Mills contractors also drew high marks for the accommodations they offered. Gildersleve's A-frame camp on Rivers Inlet featured sanitary bunkhouses and fair board. "A good camp all around," reported a union member. Mike Sahanovitch's seven-man float camp on Burke Channel consisted of only a bunkhouse and dining room, both clean and well maintained. But the lack of a dry house meant that the loggers had to hang their work clothes up to dry overnight in the bunkhouse, and they had neither a wash house nor hot water to clean up. "But those contractors up at Ocean Falls are fine fellows," commented another logger in attributing shortcomings to the pressure exerted by Pacific Mills for cheap logs.[7]

The Queen Charlotte Islands camps offered a range of living conditions. Bunkhouses at Morgan's Camp 3 on Sedgwick Bay were in "fair shape", but here too the lack of an adequate dry room made it difficult for the crew to dry clothes overnight. An "imitation cook" there served a limited menu dominated by bully beef, cold meat and sourdough pancakes. The food at Kelley's Camp 2 on Rockfish Bay was "up and down". Kelley's loggers slept in eight-man bunkhouses and enjoyed adequate washing

One view of a T.A. Kelley camp on the Queen Charlotte Islands in the 1930s. Vancouver Public Library VPL 6265

facilities, although the hot water sometimes ran out before the whole crew had cleaned up. A bullcook kept the bunkhouse tidy and changed sheets and pillowslips weekly. A chokerman earning $4.25 a day in the spring of 1936 would see his earnings diminished by charges of $1.20 for daily board and another 15 cents a day for blankets if he didn't bring his own. Allison got high marks from one logger. Conditions at Cumshewa Inlet were "of the best", ice-cream and chicken appearing regularly on the menu. The cook at a nearby A-frame camp was "not so good", but at least Allison employed no "highball or other high-pressure stuff".[8]

Safety and medical facilities also ranked high among the concerns of loggers at the isolated north-coast camps in 1936. Loggers at Morgan's Camp 3 had to fabricate a stretcher out of axes and saws to carry an injured man away from the scene of an accident that summer, the owner having failed to provide the items required under Compensation Act regulations. When a worker at the Carstairs Camp broke his leg while doing some Sunday maintenance work on a gas donkey a few miles from the camp, he waited the entire day on the beach until the boat made its scheduled return.[9]

One of the most notorious cases of medical neglect occurred in the summer of 1937 on Rivers Inlet, site of a Pacific Mills contractor operation. On the morning of July 20 a bucker cut his leg seriously and walked the half-mile to camp. A search for antiseptic and bandages proved unsuccessful, so the wife of a crewman used a pillowslip to bandage the leg. A party then took the man by boat to the Margaret Bay Cannery, a three-hour trip. But the foreman there, according to the account provided to the union by a Rivers Inlet fisherman, refused him treatment, saying the first-aid man was busy canning fish. Not until 10 p.m. did the logger finally receive treatment, after a fisherman took him to the Brunswick Bay Cannery hospital. The entire episode troubled the LSWU, especially the $2.35 monthly hospital fee that the contractor charged his employees. "No equipment, no medicine, no bandages, no first-aid man, twelve and a half hours from the

Another view of the same Kelley camp. Vancouver Public Library VPL 6265

nearest place of treatment, yet charged for hospital," said the union. "Is this a 'racket'?"[10]

Morgan and Kelley hired a doctor and outfitted a rudimentary hospital in 1937 to provide care for the men in their camps, each contributing $1.00 a month to offset the company's expense. That arrangement enabled a man stricken with acute appencicitis to receive treatment, but the doctor later expressed dissatisfaction with his $120.00 monthly salary. That summer the loggers supported his request for all of the $260.00 in fees collected each month. He succeeded in securing a salary increase, but did not return in 1938, a year that produced several serious injuries and at least one accidental death at Morgan's operations. The hospital at Queen Charlotte City provided the nearest source of treatment, a dangerous and painful journey for the injured. And there was no guarantee of treatment there either, as an unfortunate hooktender at one of Allison's camps discovered in 1938. Hit by a cable when a strap broke on the spar tree, he suffered a broken arm, collarbone and head injuries. A rushed trip to Queen Charlotte City followed, only to discover the doctor absent. After a nurse bandaged the head wound, the man was shipped to the Prince Rupert hospital, but 82 hours elapsed before a doctor there attended to the injuries.[11]

Although not directly rooted in the employment relationship, several aspects of the long distance that separated the north-coast camps from Vancouver disturbed loggers. First, they bore the entire expense of the steamship journey to and from camp. First-class passage on a CN steamship to the Queen Charlotte Islands cost $25.00, amounting to a week's wages for a chokerman in 1936. Travelling second-class set the logger back $13.50, but one had to endure a cramped and unpleasant trip in the "Glory Hole". One logger related his dissatisfaction after a trip down from the Charlottes that spring, calling for a Health Department investigation. The company jammed 43 bunks in the 36 x 21 x 8 foot space on the *Prince John*, and while only 23 were occupied the "absolute lack of ventilation" made for stagnant air conditions. Opening the portholes would have provided a

SS *Prince John* at the Skidegate oil works, 1909.
BC Archives A-07662

simple solution, but the crew denied these requests. Neither did the food, at 50 cents a meal, prompt delight. A serving of "rotten mulligan and poor watery coffee" on the first day generated a protest by the second-class passengers, gaining them better fare for the remainder of the voyage.[12]

Another report that autumn highlighted the shoddy treatment afforded loggers and other resource workers on the *Prince John*. CN provided "fine" service on the *Prince Charles* during the summer tourist season, the correspondent observed, but gave "no consideration to the comforts" of the loggers, cannery workers and fishermen who dominated the fall and winter passenger list. On a recent trip south the *Prince John*, already carrying over 50 Chinese cannery workers in second class, picked up a number of loggers at the Kelley and Morgan camps. That brought the number of potential "Glory Hole" passengers to 65, well over its already cramped capacity. Some slept in the cargo hold, while others stretched out in any available space. When some loggers asked if their meals could be served somewhere other than the "Glory Hole", the captain invited them to dine in the first-class saloon, but only upon payment of first-class fares. Travel on CN ships often meant taking a chance on finding a place to sleep, the logger concluded, "and 10 chances to 1 you take your meals on a makeshift table of old planks."[13]

Inadequate refrigeration facilities on the coastal steamships also dictated that the meals served at camp, always a real concern to loggers, bred discontent. Regardless of a cook's abilities, meat and other fresh food deteriorated during the voyage from Vancouver supply houses. "In fact the meat and other fresh supplies are almost uneatable by the time they reach camp," one of Morgan's loggers complained. Infrequent servings of fresh fruit and vegetables at some camps also drew complaint, although loggers were generous in the praise of cooks who applied their skills in turning out tasty meals. Such a man would attract loggers to particular camps and help retain their services, whereas a sloppy or unimaginative cook prompted their quick departure. In the extreme, a crew might simply demand that a cook leave on the next boat.[14]

It is difficult to judge the enthusiasm with which north coast loggers greeted the LSWU's 1936 strike call. A strike vote at Morgan's Camp 2 failed to muster enough support for a walkout, but the operator conceded an unspecified wage increase and promised not to discriminate against union members. Morgan would not recognize the union, however, unless forced to do so by an industry-wide capitulation. Circumstances were more complicated at the Carstairs camp on South Bentinck Arm, where the crew had to deal with both the foreman on site and Pacific Mills management. When the strike call came the firm's logging manager was at the camp, and denied requests for a raise. The majority of the crew then voted to stay on the job, but after the strike "down below" led to wage gains, the crew renewed its demand for a 50 cent increase. By this time the Pacific Mills superintendent had departed, and the foreman persuaded the crew to delay any action until his return. The LSWU delegate at the camp regretted this successful bluff, but a subsequent threat to strike produced a 25-cent raise for those in the lower pay levels.[15]

The 1936 strike produced uneven results along the north coast, then, and fell short of breaking the operators' control over the camps. The wage increase represented a substantial victory for the LSWU in its struggle to gain the allegiance of woodworkers, but spreading the word to the many islands, bays and inlets that stretched up the rugged coastline to the Queen Charlotte Islands presented a formidable challenge. Most of these operations shifted about from year to year, bypassed by the coastal steamships. Clearly, the union needed a boat to bring organizers into contact with the up-coast loggers, to speak to the men, distribute literature, and sign up new members. That August, Local 2783, the "Loggers' Local", purchased a 45-foot gas boat, the *Laur Wayne*, to make the inaugural cruise of what would come to be known as the Loggers' Navy. On August 31 1936 the boat left Vancouver with Local 2783 president Arne Johnson in charge. The vessel would visit a few camps in the vicinity of Vancouver and then make the voyage to the Queen Charlottes, taking collections on the way to meet the boat payments.[16]

The *Laur Wayne* proved an immediate hit with the loggers. Johnson gave those at a Sointula camp a short ride before holding a meeting, collecting $92 in donations. By the evening of September 7 Johnson had tied up at the Butedale cannery on the north end of Princess Royal Island. He had discovered that the vessel rolled alarmingly in rough weather, but apart from running into some logs on Millbank Sound, he had encountered no trouble. He proceeded to Beaver Passage, about 20 miles below Prince

Rupert before making the crossing to the Queen Charlottes. On the advice of a fisherman he crossed at night in hopes of finding calmer seas on Hecate Strait, but the winds picked up. They almost lost the lifeboat at one point due to the *Laur Wayne's* excessive rolling. "We had a pretty rough trip," Johnson reported. "Next time we tackle the straits we are going to be sure that it is more calm."[17]

A warm reception at the Kelley, Morgan and Allison camps made up for the difficult crossing. Two meetings at Morgan's headquarters camp produced $169.50 in donations and some new members. Thirteen loggers signed up at Kelley's Camp 2, and another $86.00 went into the boat fund. "The boys here were all for the idea of a boat," declared the LSWU delegate at Camp 2, "realizing it was the best means hit upon yet for organizing the camps around this part of the world." Even the Kelley and Morgan management seemed receptive to the organizers, allowing them to eat free of charge in the cookhouse. The stay on the Charlottes provided some lighter moments too, such as loggers at Camp 2 at first mistaking the organizers "for a couple of preachers".[18]

Cautious about the *Laur Wayne's* capacity to handle rough seas, Johnson held up at Queen Charlotte City with the trawling fleet after a storm forced them to abort one crossing. A fisherman there described the vessel as comfortably laid out, the hull "shaped on speedy lines", but he recommended the installation of a diesel engine. The Loggers' Navy should be seaworthy and fast, because "long cruising and bad stretches of water must be considered." Johnson no doubt agreed, but the *Laur Wayne* had already proved its value, the organizer achieving more in a few days on the Queen Charlottes "than could have been accomplished in months by correspondence". And more would be gained on the return trip, allowing visits to dozens of small operations "hidden away in various bays and inlets" that the union had never been able to contact.[19]

Meanwhile, the timber workers' affiliation with the craft-conscious Carpenters and Joiners proved to be less than brotherly. Treated as second-class members of their own union, district councils from British Columbia and the Pacific Northwest met in Portland in September 1936 to create the Federation of Woodworkers and vote in Harold Pritchett as president. The following July, federation members, with the exception of Puget Sound locals who decided to remain with the AFL's UBCJ, cast their lot with the dynamic CIO. Pritchett, a Canadian communist, received sufficient support to become president of the new International Woodworkers of America (IWA). British Columbia became District 1 of the IWA, with only 2,500 members out of the province's estimated 25,000 woodworkers. Clearly, much work remained to be done.[20]

An unfriendly body of labour law in Canada and British Columbia made the organizers' task much more difficult than in the United States, where the Wagner Act provided federal government support for collective bargaining. Working-class pressure forced Duff Pattullo's Liberal govern-

Allison Logging Company camp, Cumshewa, 1937.
Vancouver Public Library VPL 6272

ment to pass the province's Industrial Conciliation and Arbitration (ICA) Act in 1937, embodying a vague principle of workers' rights. But the legislation restricted bargaining to committees of employees rather than union representatives, and banned strikes during the drawn-out arbitration process. Determined to "stem the tide of militant industrial unionism that was washing over much of North America," the Pattullo administration had done little to ease the organizer's task.[21]

It might have come as some surprise, then, when the Ocean Falls and Powell River paper mills unionized without a strike in the summer of 1937. When the AFL's United Brotherhood of Pulp, Sulphite, and Paper Mill Workers (UBPSPMW) sent a representative to the region, the Powell River Company put up little obstruction. The organizer went on to Ocean Falls, and by the end of July both plants were almost fully organized. By the end of the year identical agreements had been negotiated between the companies, the UBPSPMW and the International Brotherhood of Papermakers, the latter representing the skilled newsprint machine operators. The contracts provided for wage increases, overtime pay and paid holidays, but perpetuated the differential that saw Asian workers earn a lower hourly wage than whites involved in the same occupation.[22]

Jurisdiction over the loggers employed by these firms remained with the IWA, which sent the *Laur Wayne* on another voyage that spring. Once again Johnson received a friendly reception from the men, returning with sufficient funds for the local – now Local 1-71 of the IWA – to pay off the boat. The local then purchased a 38-horsepower diesel engine, staging a dance and draw in Vancouver on August 25 to raise funds. When Johnson took the *Laur Wayne* north that September the craft was capable of over nine knots, double its former speed. That trip went uneventfully, although heavy seas made for another rough crossing of Hecate Strait.[23]

A. P. Allison (left)
with K. Boswell at
Cumshewa, 1940.
BC Archives NA-07112

Donations to the boat fund totalled over $200, and Johnson reported no opposition from the managers of the Queen Charlotte Island camps in holding meetings. But maintaining functioning camp committees in the face of high labour transiency would be a difficult challenge in these camps. Only at Allison's did Johnson find a committee in place and delegates on the job signing up new members, selling the *Lumber Worker* and collecting dues. Conditions at Allison's railroad camp, where flower boxes stood in front of the bunkhouses, were "as good as any and better than most", Johnson remarked. Where hostile operators and superintendents refused to allow the *Laur Wayne* to tie up, Johnson and his successors would anchor a few yards off the dock, toot the foghorn to alert the crew of their presence, and speak from the roof of the pilothouse. "No camp large or small will be missed," Local 1-71 officers assured up-coast loggers. "The Navy will call in at every camp we can locate."[24]

But the IWA struggled over the next two years, its resources drained by a bitter and unsuccessful 11-month strike against the Pacific Lime Company at Blubber Bay in 1938. High unemployment during this period also took its toll, there was the B.C. Loggers' Association blacklist to contend with, and the employer-friendly provincial labour code provided no aid to organizers. When three of Allison's loggers went to his Vancouver office in January 1938 to inquire about going back to camp after the Christmas shutdown, the operator cited their union activities in refusing to rehire them.

The *Annart* of the IWA's "Loggers' Navy".
Rare Books & Special Collections, UBC Library 1532/776/1

The IWA protested this discrimination to the provincial Labour Department, but because logging fell into the category of "seasonal" employment
the three were not considered Allison's employees at the time, and thus
had no claim under the ICA Act.[25]

Most of the camps in Local 1-71's jurisdiction had still not reopened
after the Christmas break when the *Laur Wayne* set out that May. On the
Queen Charlottes, Morgan and Kelley employed only skeleton crews at
reduced wages. Hundreds of unemployed up-coast loggers had allowed
their memberships to lapse, and by the summer of 1938 the IWA had a
mere 226 dues-paying members, most at Cowichan Lake. After this, "the
toughest year of the organizational struggle" according to Myrtle Bergren,
the IWA decided to focus its energies on Cowichan Lake and the Queen
Charlottes. The isolation and climate of the latter area made it difficult for
operators to attract loggers, and the expense of rafting logs long distances
added a cost pressure that made efficient performance by a stable crew
essential to profitable operation. The Queen Charlottes would be one of
the IWA's "operational centres", then, an isolated scene of heavy industry
where operators had an exceptional interest in maintaining worker productivity during the logging season.[26]

Over the next couple of years industry recovery and determined work
by the union's small but dedicated band of organizers enabled the IWA to

rebuild. Membership in British Columbia stood at about 1,800 by the end of 1939. Like other Communist-led unions the IWA took a militant stance after the signing of the 1939 non-aggression pact by Hitler and Stalin. District Council 1 voted to launch a campaign for a wage increase under the banner of "Boost Pay a Buck a Day" at its 1940 Convention. By this time John McCuish had taken over command of the *Laur Wayne*, playing a pivotal role in a 50-cent increase secured by Queen Charlotte Islands crews that spring. By August over 800 loggers inhabited the eight camps there, all supporting McCuish's campaign for a joint working agreement with the major operators covering collective bargaining, seniority rights, wage increases, overtime pay and safety concerns. Gains on most of the fundamental issues would not be achieved for another three years, but by the time McCuish left in November another 50-cent wage boost had been negotiated. A chokerman who earned $4.00 for a day's work in 1936 drew $5.00 or more on the Charlottes by the end of 1940, and camp committees had been established at the Morgan, Kelley, Allison and Pacific Mills operations.[27]

Only two things troubled McCuish on his return trip to Vancouver that November. The achievements on the Charlottes had been made at the neglect of the mainland camps, and after about a dozen round-trips up coast no further doubt existed about the *Laur Wayne*'s unsuitability for the Hecate Strait crossing. That December Local 1-71 solved both problems by purchasing the 40-foot *Annart*, a craft better able to cope with rough seas encountered in servicing its Queen Charlotte Islands sub-local. The *Laur Wayne* would be released for travel along the more protected inside waters of the mainland coast and eastern Vancouver Island.[28]

A major policy shift for communist-led unions such as the IWA occurred after Hitler's armies attacked Russia in 1941. The District 1 leadership swung around to support the war effort, adopting a no-strike pledge under the "Production for Victory" slogan. That would contribute to conflict between left-wing and "white bloc" IWA members, and Pritchett had also come under pressure from the CIO and U.S. government. He had to resign as president of the International in 1941 when the U.S. Immigration Service denied him entry to attend the union's convention, citing his Communist Party membership. But the factional infighting between reds and anti-communists that disrupted organizing efforts on the Lower Mainland and Vancouver Island does not appear to have infected the Queen Charlotte Islands.[29]

McCuish set out for the Charlottes on the *Annart* in early 1941 seeking another $1.00 wage increase and a union agreement. Local 1-71 also wanted the operators to assume the cost of travel – to pay a one-way fare after three months of work and return fare after six months. But reports of intimidation by some of the bosses circulated, suggesting that any management tolerance of union activity was at an end. An Allison Logging Company manager went so far as to threaten an IWA business agent who sought entry to his camp in April. Mackenzie King's Liberal government had adopted Order-in-Council PC 2685 the previous June in hopes of curtailing industrial unrest that might undermine the war effort. That order encouraged but did not compel employers to recognize unions voluntarily, negotiate, and submit disputes to a conciliation process.[30]

Labour, already chafing under federal wage controls, found further cause for discontent with the order. But the document did provide mild support for workers' democratic rights, sufficient to justify an IWA protest to the federal Department of Labour. District Council secretary Nigel Morgan cited the gains made in the previous year as the Queen Charlotte Island operators responded collectively to the union's demands. At least 700 of nearly 1000 loggers on the islands belonged to the IWA, Morgan asserted, and the union's vessels required free access to the camps for the purpose of organizing, collecting dues and handling grievances.[31]

Moreover, he stressed, in the past the union's representative had visited bunkhouses after working hours, a practice that threatened no disruption of production. Allison made it clear to IWA representatives in Vancouver that access would no longer be permitted, despite PC 2685. Did the order have any meaning at all, Morgan asked? The answer, according to Laura Sefton MacDowell, is that it did not. Neither employers nor the government itself adopted its principles until the adoption of stronger legislation toward the end of the war. For the moment, the union could only condemn the Allison company's "Hitler-like attitude" of refusing employees the right to meet their elected officials. In a sample of the rhetoric that would accompany the union's "wartime campaign", the IWA pointed out the contradiction of the Allied fight for liberty in the absence of democracy on the home front. "While basking in the super profits procured through government war orders," lumber barons like Allison pursued a "concentration camp policy likened only to that of Germany and Italy."[32]

Allison's hard-line defence of the open shop notwithstanding, the Queen Charlotte loggers secured another 50-cent raise in August 1941. Allison even seemed to soften his stance on union activity, allowing meetings in the cookhouse. That October Local 1-71 decided to take advantage of the momentum. At Cowichan Lake the loggers at the Lake Logging Company had just achieved a ground-breaking agreement that afforded Local 1-80 the closed shop, seniority and leave-of-absence rights. The Loggers' Local would try to gain the same benefits from the Queen Charlotte operators, electing negotiating committees at the camps, where the union now claimed 90 per cent of the loggers as members. The representatives began by holding a conference which produced a proposed collective agreement. After discussion at the camps the document underwent revisions for presentation to the operators. The IWA appealed to Ottawa for enforcement of PC 2685, and Pacific Mills logging manager P.S. Bonney indicated that firm's willingness to begin negotiating an agreement after a federal Department of Labour representative had met with executives.

Thomas Kelley expressed a similar sentiment. But if the union expected the process to go smoothly, it found out otherwise as the fight for a Queen Charlotte Islands agreement dragged on until late 1943.[33]

A mass meeting of the Queen Charlotte Island loggers at Vancouver's Hastings Auditorium during the 1941–42 Christmas shutdown set the stage for discussions. After a negotiating session between Pritchett and District Vice-President Haljmar Bergren with Thomas Kelley fell through, the federal Department of Labour arranged a late-January meeting between IWA officials and representatives of the Allison, Kelley and Morgan companies, and Pacific Mills. The Joint Operators Committee asked for a 10-day adjournment to review the proposed agreement, but had not yet responded by late February. "Seven months passed in such shilly-shallying," IWA lawyer John Stanton recalled. Morgan and Kelley now followed Allison in adopting what the IWA called "Hitler-like anti-labour policies". Notices posted at their camps denied union representatives access to company property.[34]

The IWA, in contrast, pledged to serve as a willing partner of management and government in the fight against fascism. The union even dropped its demand for recognition for the time being in exchange for an agreement giving workers the right to elect bargaining representatives "at large". The organization would settle for a pact on the more immediate issues, work with the employers to encourage skilled loggers to go to the Queen Charlottes, and participate in joint management-labour-government industrial councils at the camps to boost spruce production and settle minor grievances. As the operators' silence carried on into March, the IWA charged them with deliberate sabotage of the war effort. "Fed up," men were leaving the spruce camps.[35]

The union had no desire to become entangled with the federal government's Industrial Disputes Investigation (IDI) Act, which early 20th-century legislation extended at the outset of the war to cover most industries. The law imposed a "long, drawn out" conciliation process before workers could legally strike, a delay that gave employers time to counter union offensives with persuasion, the offer of management-dominated "employees' committees" or the recruitment of strike-breakers. But with no alternative, the IWA went ahead with strike votes in the Queen Charlottes that spring as a prerequisite to applying for a Conciliation Board under the IDI Act. An overwhelming expression of support for strike action set in motion a last-minute series of negotiations between union officials, the companies and federal officials concerned about spruce production.[36]

Attributing problems in spruce output to the refusal of the Morgan, Kelley, Allison and Pacific Mills organizations to bargain collectively with the union chosen by 90 per cent of their employees, the IWA set out five key demands. First, the operators should enter into an agreement with the union embodying grievance committee procedures, safety committees, payment of wages in legal money orders, seniority and leave-of-absence provisions, and freedom from discrimination for union affiliation. Industrial councils of one government, three management and three labour representatives should be established in each camp to promote efficient allocation of workers, introduce procedures for the training of inexperienced men and act as a binding court of last resort in dealing with disputes.[37]

To encourage loggers to return to the Charlottes, the union proposed that the companies pay the one-way steamship fare after three months of work, both ways for workers who had worked at least six months. Refrigeration for perishable foodstuffs should be provided both on the steamships and at the camps. Improved medical care also figured into the IWA's submission, involving a demand for aircraft transport to Vancouver for injured loggers and improved treatment facilities at the camps. Finally, the union requested delivery of mail by air to the Queen Charlotte Islands camps and paid vacations to break the monotony of work in isolated camps. Federal officials responded by saying that they had "forcibly impressed the logging operators with the absolute necessity of getting production up to requirements," urging similar cooperation from labour.[38]

IWA District 1 officials achieved a settlement of most of these demands thanks to the intervention of National Selective Service and Timber Control officials, but negotiations on a collective agreement and union recognition broke down in the face of the operators' commitment to the open shop. Still, Pritchett and District secretary Nigel Morgan negotiated extension of the terms to a new Pacific Mills camp on South Bentinck Arm, and that September a small six-bed hospital equipped with an X-ray machine and operating table opened at Cumshewa Inlet. Federal officials authorized a 30 per cent wage bonus to Queen Charlotte Island loggers who stayed on the job for over a hundred days that autumn in response to the urgent need for aircraft material. Spruce loggers would also receive a deferment from military service. On the other hand, the Union Steamships Company had not yet fulfilled its commitment to install adequate refrigeration units on its ships.[39]

But on the central issue, the four Queen Charlotte operators formally opposed granting the IWA recognition. Their stated reason, initially at least, was a desire to avoid the jurisdictional disputes between unions that had caused strikes in the United States. Since many of their logs went to pulp-and-paper plants organized by other unions, they foresaw the same result for their operations. They would enter into agreements directly with their employees, but not the IWA Finally, in October 1942, the federal government announced the establishment of a non-binding conciliation board to investigate the year-old dispute. Cooperative Commonwealth Federation (CCF) MLA Arthur Turner would represent the union, with Vancouver lawyer R.H. Tupper representing the companies under chairman Judge A.M. Harper. At issue was whether the companies should recognize the IWA as the sole bargaining agent for the Queen Charlotte loggers, after reaching a basic agreement on wages, seniority and vacations.[40]

Hoping to avoid both the expense of bringing witnesses down from the Charlottes and an interruption of spruce production, the board persuaded the two sides to meet in advance of hearings. It looked for a time in late 1942 as if a deal had been worked out, but at the last minute the companies' lawyer, C.H. Locke, communicated their refusal to sign an agreement with the IWA. The hearings would go ahead, then, involving nine sittings between January and late May 1943. Locke hammered away at a single theme throughout the proceedings: the operators' willingness to deal directly with their employees but not the union's communist leadership who had failed to support the war effort with the proper patriotic fervour during the time of the Hitler-Stalin pact.[41]

In the interim, Local 1-71 president John McCuish kept busy on the Charlottes, where Aero Timber Products had taken over Allison's operations. As a Crown corporation Aero fell outside the existing dispute, but took the same position as the other companies in the region. Thus, when McCuish arrived on the *Annart* at Cumshewa Inlet on a Saturday that March, management informed the crew that they could meet in the cook-house on the condition that no "outsider" attend. In that event, the camp committee told the superintendent on Sunday morning, the crew would hold a special meeting on Monday, a work day. Aero wilted in the face of the loggers' defence of their representative; the meeting went ahead with McCuish as "guest of honour". Proceeding to Aero's Masset Inlet camps, McCuish helped press Aero management to secure a doctor and hospital for the area.[42]

The Queen Charlotte operators must have rejoiced when the regional oil controller informed Local 1-71 that its fuel permit would expire on April 1 1943, threatening future purchases for the *Annart* and *Laur Wayne*. The IWA estimated that its "navy" travelled an average of 13,000 miles a year by this time, bringing the organization into contact with 4,000 loggers in about 90 camps. The protest reached sufficient volume to secure permits for additional supplies, enabling District President Pritchett to accompany McCuish on a speaking tour of the Charlottes in April and May. He addressed 21 meetings while on the islands, his first voyage on the *Annart* coinciding with Local 1-71's final payment to North Vancouver's Roger Prentice, whose loan had made the vessel's purchase possible.[43]

During this period both federal and provincial governments were forced to respond to the wave of labour militancy that threatened the war effort and increased support for the socialist CCF. Ottawa's mishandling of a strike by gold miners at Kirkland Lake, Ontario for union recognition in the winter of 1941–42 had infuriated the labour movement. Trade union support almost brought the CCF to power in Ontario in August 1943, a clear warning sign to governments across Canada. Mackenzie King had already begun moving hesitantly to retain organized labour's support, extending the principle of collective bargaining to those employed by Crown corporations in December 1942. But within weeks some 13,000

Aero Timber Products' Juskatla camp, 1945
BC Archives NA-08439

Ontario and Nova Scotia steel workers struck, pushing King further down
the road toward a new labour law that would grant workers' demands for
collective bargaining rights.[44]

British Columbia's coalition government, formed by Conservatives
and Liberals early in 1942 to hold off the CCF, also felt pressure from
organized labour to amend the ICA Act. Premier John Hart responded in
March 1943, providing for government certification of unions as bargaining agents if membership reached a majority of a firm's employees. The
legislation, in addition to supporting compulsory recognition, made it illegal for bosses to interfere with organizers or encourage company unions.
IWA officials hailed the act as a "Bill of Rights" for labour, and in April the
union secured the industry's first agreement with the Batco Logging Company at Campbell River. Pritchet's stay on the Queen Charlottes culminated with a conference at Aero's Juskatla Inlet headquarters with 27 delegates
representing the firm's camps. The meeting adopted an agreement based
on the Batco contract, elected a negotiating committee, and informed Aero
head Robert Filberg of their desire to open negotiations immediately. At
the same time Local 1-71 applied for certification as the bargaining agency
for the 530 loggers employed by Morgan, Kelley and Pacific Mills.[45]

Meanwhile, the federal arbitration board hearings were winding down.
John Stanton concluded the IWA's case on May 19 1943, stressing the centrality of unions in any concept of collective bargaining, recent provincial
and federal legislation supporting union agreements, and the organization's
support of the war effort. The board's majority decision, handed down in
early June, favoured the union in every respect. Chairman Harper and
Arthur Turner concluded that "the fears the employers entertain of what

Pacific Mills' loggers, Queen Charlotte Islands, 1945.
City of Vancouver Archives CVA 586-3707

would happen if they deal with the officials of this union are unfounded," and the companies should enter into a one-year agreement with Local 1-71. "In our opinion the good sense and sound judgement of Canadian work- men can be trusted," they explained, condemning "paternalism in the selection of representatives of labour". Tupper offered a contrary opinion in line with the operators' position, but the *B.C. Lumber Worker* described the majority report as "the most outstanding and significant victory for the IWA in British Columbia."[46]

By this time IWA locals had been certified under the ICA Act as bar- gaining agents for workers at several large Vancouver Island operations, along with Aero Timber on the Queen Charlottes. Nothing, it seemed, could stop the union's momentum now. But the entire coastal industry, in the form of the B.C. Loggers' Association, was determined to stall the union drive on the Queen Charlottes. Later in June, Pritchett and Nigel Morgan met with operators' negotiator R.V. Stuart and George Curry of the federal Department of Labour to ask that the award be carried out. Fail- ing a settlement the union could apply to the federal agency for a strike vote, an outcome that seemed increasingly possible as subsequent meetings produced consent to the terms of an agreement but no willingness on the part of the operators to recognize the IWA. All eyes were on the dispute, now dragging into its second year, when in July 1943, 42 firms signed a public statement expressing support for the Queen Charlotte operators in refusing to enter into an agreement with Local 1-71.[47]

The Union Steamship *Camosun III* at Prince Rupert, ca 1948.
BC Archives B-04345

IWA leaders, for their part, wanted no work stoppage that would violate the no-strike pledge and undermine the union's war record. But the
Queen Charlotte companies, now in open defiance of both the federal conciliation award and the provincial ICA Act, refused to budge. Accordingly,
on July 15 the union went ahead with a request for a strike vote on the
Charlottes. Aero general manager Robert Filberg, with less freedom of
action as the head of a Crown corporation under federal law, indicated his
willingness to negotiate with the IWA if given proof that the union represented the majority of the employees.[48]

While Local 1-71 proceeded to gather that evidence, loggers on the
Charlottes grew impatient at the slow pace of negotiations. Although
Aero's takeover of Allison's operation had led to better conditions, including spring-filled mattresses, Morgan's camps remained a cause of discontent. Coal oil lamps provided dim lighting in the bunkhouses at his Huxley
Island operation. In the absence of a dry house, refused by management,
"the smell and stench from sweaty and wet clothes is awful," one logger
reported. Lack of adequate refrigeration on Union Steamship boats coming to the Charlottes continued to generate complaint. One shipment of
meat from Vancouver arrived at Cumshewa Inlet at "close to spoiling
point". Another from Prince Rupert was simply loaded onto the deck of the
ship for passage to Masset Inlet, according to a report. The lack of premium pay for overtime, recently granted to sawmill workers, constituted
another grievance for loggers who routinely worked 10-hour days.[49]

The cookhouse at Pacific Mills, Queen Charlotte Islands, 1945. City of Vancouver Archives CVA 586-3645

Facing an IWA deadline of September 8 for action on the strike vote, federal Department of Labour representative George Currie brought union and company officials together in late August to discuss a one-year agreement he had drawn up. The package, which included union recognition, free transportation, seniority and overtime pay after eight hours, gained IWA acceptance. But the operators declined the settlement proposal, demanding that the provincial Labour Department conduct a new vote to determine if the Queen Charlotte loggers wished the IWA to serve as their bargaining agent. Under the ICA Act such a vote could be held every six months, a provision that would draw the IWA into a never-ending cycle of votes to maintain its bargaining agent status.[50]

Refusing to entertain that prospect, district leaders set a strike vote for September 30. Work would stop the next day, bringing the flow of spruce to a halt. The District Council declared September 27 "Queen Charlotte Islands Support Day", asking woodworkers to donate a day's wages to a strike fund. Faced with that threat, the federal Minister of Labour Humphrey Mitchell appointed Winnipeg jurist Justice S.E. Richards to conduct an immediate inquiry. The vote would go ahead as scheduled, but Local 1-71 and District officials recommended that the Kelley, Morgan and Pacific Mills loggers delay any action for a week, until October 7. Local 1-71 had already applied for federal certification as the bargaining agent for Aero loggers under the special regulations set up for Crown corporations. The secret ballot produced a 92 per cent vote in favour of a strike. Meanwhile, up to 30 provincial arbitration boards at other coastal operations awaited the outcome of what shaped up as a pivotal event in the history of the provincial labour movement.[51]

In the end, some 90 loggers at the Skedans Bay operation of the Kelley Logging Company took matters into their own hands, striking on October 6. If meetings with Richards failed, the district would have no choice but to authorize a full-scale walkout on the Charlottes to compel union recognition. When the negotiations under Richards broke down as expected, the strike went into effect at midnight on October 7. "The IWA has established an all-time record for patience in Canadian labour relations in this dispute," Nigel Morgan explained. Despite its best efforts to avoid "any situation which might ... impair the war effort", the union had no choice but to honour the will of the membership. Over 500 loggers heeded the call, triggering a two-week work stoppage at the eight Kelley, Morgan and Pacific Mills operations. "It was the first real tight strike where nobody scabbed," John McCuish recalled. Cooks had arranged to have plenty of food on hand, and the men remained in the camps. Deer meat provided by hunting parties supplemented cookhouse stocks. Funds from IWA locals and other unions enabled crews to continue paying room and board charges to the companies.[52]

Stephen Gray argues that the close links between the Queen Charlotte operators and the coastal pulp-and-paper plants contributed to the strike's quick resolution. Aero Timber's operations, geared to aircraft production needs, continued throughout the strike as the IWA pursued certification under federal law. But Pacific Mills at Ocean Falls relied on the Charlottes for 70 per cent of its log supply. The Charlottes also provided Powell River with its only source of spruce. Neither of these giant enterprises could afford a prolonged disruption of production. Kelley Logging, the Powell River Company's subsidiary, was the first to bend under the weight of union solidarity. According to John Stanton, Kelley informed Pacific Mills and Morgan of his intention to sign an agreement alone if they intended to "fight to the last drop of blood". But Kelley didn't have to go it alone; the others "caved in" as well on October 23.[53]

The one-year agreement signed on November 1 recognized Local 1-71 as sole bargaining agent for the workers in all camps where the union received provincial Labour Department certification. It extended seniority rights after 60 days of employment, provided for leaves of absence in the event of illness or injury, afforded status to safety committees, and established procedures for settlement of grievances. The existing wage scale would remain in effect for the term of the agreement, the companies agreeing to twice-monthly paydays. Board prices also went unchanged, but were subject to negotiation. The eight-hour day came into effect on a 48-hour week, with time-and-a-half for overtime. For its part, the union agreed to a no-strike pledge on the Charlottes.[54]

The union's expectation that the Queen Charlotte agreement would set the pattern for subsequent negotiations was fulfilled in the coming

The married quarters at Pacific Mills' Sandspit camp, 1945.
BC Archives NA-08436

weeks. Within two days the Lake Logging Company at Cowichan Lake signed a similar accord. Aero Timber Products followed soon after with an agreement covering the company's 850 loggers. Arbitration boards along the coast district resumed negotiations, leading to discussions between the Loggers' Association and District 1 officials for a coast-wide master contract. "After some 80 years the 'Open Shop' has been broken," the IWA announced in December, after signing a document with 20 of the largest coastal operators that placed 8,000 to 10,000 woodworkers under a single collective agreement.[55]

While the IWA continued its organizing drive, across the country workers were becoming increasingly unhappy with the federal government's wartime wage control policy. Facing strikes in critical sectors of the economy as the war dragged on, and the ominous growth in support for the CCF, Mackenzie King finally acted to maintain political and social order. Order-in-Council PC 1003, enacted in the spring of 1944, guaranteed labour's right to organize and bargain collectively. The measure, based loosely on the American Wagner Act, banned company unions, confirmed the right to strike in the absence of a collective agreement, and contained the "distinctly Canadian" principle of compulsory conciliation prior to a legal strike. British Columbia adopted the federal labour code in April for the duration of the war, allowing the IWA to take advantage of less onerous certification provisions to increase membership among coastal loggers.[56]

Indeed, the union pressed aggressively to extend its reach throughout 1944. In early February Local 1-71 announced the signing of its "best agreement to date" with the Gildersleve Logging Company at Rivers Inlet. Combining features of the Queen Charlottes accord and the Master Agreement, it included provisions for the check-off of union dues, paid holidays and hiring through a union hiring hall established in Vancouver. Gildersleve's small camp brought in few new members, but by May the IWA had 83 operations under the Master Agreement, covering over 15,000 workers.[57]

At the same time, camp committees gained experience in negotiating improved working and living conditions. At one Aero operation the grievance committee requested better washhouse facilities, including the installation of mirrors, more basins and a clothes wringer. At another workers approached management for a new lighting plant for the camp, sidewalk repairs, more showers, a larger dryroom and extermination of bedbugs. That summer, as Aero Timber began to wind down its operations, loggers at Port Clements voted unanimously to request the immediate removal of the cook. When management balked the committee demanded the assistance of union headquarters, leading to a visit by McCuish that produced a satisfactory resolution.[58]

The always contentious issue of meat shipments from Vancouver also continued to spark complaint that summer, when poor refrigeration and hot weather combined to produce its arrival "in a very poor state". Improved medical facilities, the establishment of libraries, and safety matters topped the agenda at meetings throughout the Queen Charlottes. Pacific Mills loggers at one float camp requested that a boat be stationed near the logging area during working hours for prompt removal of the injured. McCuish noted a significant improvement in labour relations since the camps came under union agreements, praising the work done by grievance, safety and production committees.[59]

McCuish launched an organizational campaign in the Terrace area that autumn, observing that he "hadn't seen such conditions in camp for over 20 years". That process dragged on over several years as the union sought to extend its reach into the northern interior. Consolidating wartime gains on the coast proved easier if no less contentious. A landmark strike in the summer of 1946 gained a wage increase, voluntary check-off of union dues, the 40-hour work week in summer and 10,000 new members. By the summer of 1947 major Skeena River operators such as Little, Haugland and Kerr were included in negotiations between the IWA and the Northern Interior Lumbermen's Association (NILA). Olaf Hanson dug in his heels, refusing to negotiate with the union. "There is nothing to negotiate," he told the Local 1-469 secretary at Terrace. "I pay good wages and respect the hours of the work act." Hanson's obstinacy notwithstanding, the IWA's 1947 Master Agreement with the NILA achieved union security provisions based loosely on the coastal model. Organizing the mills and camps in the interior proved a long, bitter struggle, however, producing prolonged strikes over the elusive goal of wage parity with the coast. Bitter factional fights also lay ahead, resulting in the ouster of those leftists who had established the union's foundation, but the IWA would not have scored its wartime victories without those who led the up-coast fight.[60]

Decked Colcel logs on the bank of the Kitsumkalum River, 1950.
BC Archives NA-10200

6 "Era of Error"

Sustained Yield and the Dynamics of Development, 1945–70

The two-and-a-half decades after World War II brought dramatic changes to the northwest timber economy. Healthy world wood-product markets, the arrival of multinational capital to pursue pulp-and-paper developments, and the adoption of sustained-yield policies all seemed to mark the forest industry's maturation into a job-and-wealth-producing engine of limitless growth. New paper plants at Kitimat and Prince Rupert, more intensive logging by firms shifting north from depleted south-coast holdings, and a healthier sawmilling sector along the Skeena River gave the northwest a share of the long postwar economic boom.

Expansion came at a price, however, even if the total bill would not come due for a time. Some expressed misgivings and even outright opposition as the new structure took shape. Small operators criticized the government's tendency to award control over the resource to corporate interests with the capital to build costly paper facilities. Some of the new up-coast Tree Farm Licences developed to attract investment went to firms requiring northern fibre to supply existing Lower Mainland and Vancouver Island plants. Still, the massive tenures would be operated in accordance with sustained-yield principles, benefiting the entire province. That, at least, is what corporate and policy-making elites said in response to the complaints of dissidents in places like the Bella Coola River valley. Technological changes that kept employment from rising in proportion to harvest rates could also be accepted given the overall rate of development and the pressures of a competitive world marketplace.

Arguments were also at hand to counter the protests of First Nations observing industry's consumption of forests on traditional territories. First Nations were free to compete for Crown timber, the Forest Service informed them, simply by following established timber-sale procedures. Left unsaid was the well-founded expectation that lack of capital and bureaucratic requirements would discourage such participation. When First Nations did apply, established industrial interests simply out-bid them in a process designed to uphold the Crown's right to a fair share of timber values. Of course, none of this went to the heart of First Nations' concerns

Pacific Mills' Martin Inn Residential Hotel, 1953.
BC Archives I-50634

– the despoliation of lands and resources that had never been alienated through treaty.

For the Social Credit government that administered British Columbia's forests through the 1950s and 1960s, neither aboriginal claims nor the pleas of small operators would be permitted to disrupt the process of consolidation and integration that held the key to provincial development. The forests were a source of profit and revenue needed to stimulate employment, community stability and the provision of social services. That the complacency and optimism these policies engendered rested on a thin foundation of scientific knowledge and management planning seemed inconsequential in a land of such abundance. Not until the late 1960s would the social and ecological costs of the sustained-yield project begin to gain attention. There were limits even to the northwest's resources, and the corporations would stay only so long as the cost of extraction did not exceed their commodity value.

Although the overall cut from the Prince Rupert forest district declined in 1946 due to reduced output from the Queen Charlotte Islands, rising domestic and export markets in 1947 heralded the onset of the postwar boom along the north coast. Producers of lumber, pulp and paper, poles, and ties all enjoyed higher prices, boosting production levels. By the end of 1948, returning servicemen had lifted the number of portable interior mills to well over a hundred, twice the 1944 number. They pushed into more remote areas, operating on a narrow profit margin, and when lumber

prices dipped late in 1948, District
Forester J.E. Matheson predicted
that many would be forced out of the
industry if the trend continued.
Increasing skidding distances placed
additional financial pressure on interior loggers. As readily accessible
timber disappeared, and the number
of experienced teamsters dwindled,
operators discarded horses in favour
of caterpillar tractors on their timber
sales. That trend produced a corresponding increase in damage to
residual timber from blades on the
"cats", forcing district staff to introduce penalties on contracts and
threaten careless operators with cancellation of permits.[1]

Tractor and arch haul logs near
Houston, 1951.
BC Archives NA-12303

By the end of the 1940s the
tremendous expansion in interior
lumbering, fuelled in large part by
freight rates that allowed Terrace mills to penetrate American and eastern
Canadian markets, found operators "beginning to get in each other's way".
"Realization of limited timber supplies is spreading," Marc Gormely
observed in 1950, pointing out that activity had picked up as far north as
Stewart and Atlin. The number of operating saw and planer mills had
reached 292, employing roughly 1,200 men, and some 2,200 loggers
manned the district's camps. Ocean Falls remained the hub of the coast
industry, having launched a $5 million postwar expansion program in 1945
to prepare for the anticipated boom in pulp-and-paper demand. Construction of the 250-room Martin Inn, a residential hotel for single workers,
allowed Pacific Mills to erect new family dwellings in place of several dor-
mitories. Plant improvements brought annual production of newsprint,
kraft and speciality papers to about 140,000 tons in 1947, up from the prewar total of 103,000, and profits exceeded $3 million in 1949.[2]

The postwar boom gave Ocean Falls' 2,500-plus residents stability, and
parents an opportunity to raise children with confidence in the community's future. The firm had established a forestry department in 1944, pledging to place its holdings "on the kind of managed basis that will assure adequate supplies of pulp timber for its sustained operations". The 1946
acquisition of timber licences in the Kitimat area consolidated fibre supplies for the time being, but Pacific Mills was drawing logs from an ever-
widening range of sources in order to generate annual sales of $30 million
in the early 1950s. In addition to operating camps at Sandspit on the
Queen Charlotte Islands, Beaver Cove on northern Vancouver Island and

Opening of a new school at Ocean Falls in 1952.
BC Archives I-50560

South Bentinck Arm near Ocean Falls, the firm drew over half of its fibre supply from contractors and log-market purchases.[3]

Paper manufacturing at Ocean Falls provided a strong link between work and community during the postwar years, but elsewhere on the north coast few such opportunities existed. In the interior, the "transitory and scattered" nature of logging dictated that camp conditions left "much to be desired". Insulated two-man cabins constructed on skids represented the height of comfort to be found in that region, reflecting a modest investment in crew stability. On the Queen Charlotte Islands the absence of manufacturing served to limit the potential for community development, although new corporate owners took steps to cultivate a "home guard" of skilled loggers. After buying out J.R. Morgan's Cumshewa Inlet and Huxley Island holdings and organizing a subsidiary called Northern Pulpwoods in 1946, Pacific Mills built what it advertised as a "permanent logging community" for about 175 people at Sandspit on the northeast tip of Moresby Island. There the firm rented out centrally-heated cottages with full plumbing and electricity to married men, and four-man cabins for single loggers. Amenities included a school, general store and Canadian Legion hall for dances and movies.[4]

Sandspit represented the most outstanding example of "permanent residential camps" on the Charlottes. Elsewhere Northern Pulpwoods employed a hundred loggers at Morgan's old Cumshewa Inlet operation, operating trucks at both sites to carry logs to dumping sites for bundling into Davis rafts. The Powell River Company took over the Charlottes' only rail operation at Skidegate, renaming it the Aero camp. A smaller truck

Log booms line the shores of Cousins Inlet, with Ocean Falls in the background, 1963. BC Archives I-50542

camp at Mathers Lake provided another outlet for logs destined for pulping at Powell River, and the firm's holdings adjacent to Queen Charlotte City promised to transform that community into a "booming logging town, just as it was once a thriving sawmill town." Of the former "Big Three" of Queen Charlotte logging only Tom Kelley remained active after the war, operating a truck show for the Powell River Company on Justkatla Inlet. Another stride in the corporate takeover of the islands' timber came with the 1951 merger of the Alaska Pine group with the B.C. Pulp and Paper Company. The new Alaska Pine and Cellulose organization drew its Queen Charlotte Division logging operations at Pacofi Bay and Moresby Camp into a vast structure that included pulp mills at Port Alice and Woodfibre, the former Swanson Bay pulp leases, and several Vancouver Island and Lower Mainland sawmills.[5]

The Queen Charlotte Islands, then, represented in microcosm the classic case of a British Columbia resource hinterland, its timber drained off to create profits and jobs at southern processing centres. In 1949, fully 60 per cent of the Prince Rupert District's coastal log production originated on the Charlottes. Sixty-eight per cent of the one million feet scaled that year left the district, 30 million feet went into paper production at Ocean Falls, leaving less than two per cent for local manufacture. Lumping the Queen Charlotte totals in with the cut on the mainland coast produced only a slightly less disconcerting set of figures. Together, the coastal forests yielded a sawlog cut of over 174 million feet. Ocean Falls took almost 51 million feet of this output, and only 1.4 million feet went to local sawmills.[6]

Critiquing this state of affairs, District Forester Marc Gormely pointed

Pacific Mills' Sandspit camp, 1947.
BC Archives I-22233

to "the desirability and perhaps opportunity for an increase in manufacturing plants in the Coast Section of this District." Choosing his words care-
fully, Gormely went on to suggest that the analysis "may also indicate that
the time has arrived for consideration of reserving the present supply of
timber in this District for the industries that must eventually be established
here, or at least ensuring that the 'export' of the resources of this District
are not encouraged by reduced stumpage rates." Here the district forester
referred to the policy that allowed Queen Charlotte operators to include
Davis rafting and towing costs in stumpage appraisals, a principle that lowered their costs and arguably discouraged local manufacturing. Although
the Charlottes lacked milling facilities, district staff believed that current
production levels warranted their establishment.[7]

Gormely's prescription for policies more attuned to local economic
development generated no apparent reforms. And perhaps the explanation
for lack of response lies partly in the assumption that the province had
already taken steps to remedy the northwest's problems in conjunction
with a conversion to sustained-yield forestry. Indeed, the first of the tenures
designed to achieve this advertised triumph of capital investment, sustainability, economic development, and community stability over "cut-and-
run" exploitation went to Celanese Corporation in 1947 in exchange for
the construction of a new pulp plant at Prince Rupert. The Celanese subsidiary, Columbia Cellulose, thus became the first recipient of a Forest
Management Licence (FML), involving a massive transfer of control over
forest management to private industry under the supervision of the
province's Forest Service.

The new policy's origins date back to a 1942 memorandum penned by Chief Forester C.D. Orchard in the context of growing concern over dwindling timber supplies among established operators on the south coast. In a detailed proposal to H. Wells Gray, Minister of Lands in British Columbia's Liberal-Conservative Coalition government, Orchard proposed a new system of "Forest Working Circles" that would see industry's private land and old temporary tenures pooled with adjacent Crown forest in units of sufficient size to support operations in perpetuity. Company and government foresters would determine annual harvests on the units, with the Forest Service exercising the right of working plan approval and inspection. Such perpetual tenures would correct the flaw of existing arrangements that confined operators' interests to timber liquidation, encourage sustained-yield planning, stabilize existing operations, and promote the establishment of new enterprises. Orchard, in short would use the incentive of perpetual tenure to bring industry behaviour in line with the public interest by permitting holders to "see the possibility at some later date of retrieving capital invested and profits delayed in the immediate interest of forest conservation and perpetuation."[8]

Orchard's proposal set the stage for the 1944 Royal Commission under Chief Justice Gordon Sloan, an exercise considered essential to prepare the public for the new tenure policy that would remove much of the province's best remaining forestland from competitive bidding under the timber-sale program. Small operators took an immediate dislike to Orchard's plan, rightly suspecting that it would favour large enterprises with the capital to invest in new manufacturing facilities and forestry staffs. But it seems clear that public officials and most corporate executives shared a consensus that Orchard's proposal provided the genesis of a workable tenure system capable of attracting investment capital needed for a new generation of pulp-and-paper plants. Firms would benefit from secure tenure over Crown forestland in dealing with lenders, and the state would turn the responsibility and expense of management over to the holders. Mill expansion and construction would boost employment, stabilize communities, and place the province on the sustained-yield path. Much bickering would take place over the precise details of the industry-state relationship on the new tenures, but the course had been charted.[9]

As Sloan gathered evidence and prepared his report, established operators and new arrivals such as the Bentley and Koerner families competed fiercely for timber, buying up existing operations and investing in private land and old timber licences. By the time Sloan submitted his report in January 1945, industry was poised to take advantage of the anticipated boom in the demand for wood products that peace would bring. Orchard expressed satisfaction that Sloan's proposals embodied the principles of his 1942 proposal. Qualified operators should be permitted to retain their existing old tenures in perpetuity after logging, provided that they maintained their productivity and cut on a sustained-yield basis. Since none possessed supplies adequate to achieve such yields at existing production levels, successful applicants should be allocated sufficient Crown timber to maintain operation while cutover licensed and private land restocked, and the new forest grew to commercial dimensions. Thus, on the second and subsequent rotations, the entire working circle would provide the supplies necessary to permit sustained operation. Or so Orchard assumed.[10]

The government introduced its sustained-yield legislation in February 1947, proposing to give FML holders sole use of Crown land for the purpose of growing and harvesting successive timber crops. After several amendments to address industry concerns, the new Forest Act became law that April. A couple of firms were already involved in negotiations by this time. The Celanese Corporation – with no prior involvement in the province and no timber holdings to contribute – headed the line. The firm's plans were ambitious, involving construction of a $25 million, 250-ton capacity plant at Port Edward on Watson Island near Prince Rupert to produce cellulose for the manufacture of yarns and fabrics at the company's American plants. This would be an entirely new industry, then, involving no competition with existing pulp-and-paper manufacturers in the province.[11]

News of the forthcoming award of the FML, covering unspecified lands in the watersheds of the Nass and Skeena rivers, drew a mixed response in the region. A Smithers real estate agent declared: "Anything good for Prince Rupert is good for us. Heavy industry there will insure a stable prosperity for the district." Others, especially existing regional logging and sawmill operators, were somewhat less bullish. "The new Celanese plant has been given a sort of 'control' over all forest resources between Port Edward and Hazelton," said Duncan Kerr of Little, Haugland and Kerr in Terrace. The *Cariboo and Northern B.C. Digest* predicted that implementation of the Sloan report would "end in the complete freezing-out of small operators throughout the province, while large corporate enterprise is to be handed on a platter, the forest resources of B.C." The Celanese investment would bring a certain degree of prosperity to the region, logging contractor and editor Mike Sahonovitch conceded, but only by turning over "managership" of billions of feet of Crown timber to American interests. "Future generations of Canadians can have what is left in the area after the ... Celanese Corporation has first picked out the 'cream of the crop' for its own use," Sahonovitch predicted.[12]

The well-founded concerns of small operators aside, local business interests greeted news of the Prince Rupert pulp mill with unreserved optimism. Award of the FML to Celanese's new subsidiary Columbia Cellulose (Colcel) in 1948, involving 17 different tracts covering 668,000 acres in the watersheds of the Nass and Skeena rivers and several coastal inlets near Prince Rupert, seemed an acceptable price for a giant new enterprise. Prince Rupert's population had dwindled from its wartime high of 25,000 to some 15,000 by 1948. Now, civic officials looked forward to doubling

Columbia Cellulose plant at Port Edward 1953.
BC Archives I-28907

that total as the town made its long-awaited transition to "thriving industrial and shipping metropolis". Nor would the transformation be limited to Prince Rupert, where 500 mill workers would make their homes. A similar number of logging jobs would be centred around Terrace, headquarters of Colcel's harvesting operations. Farmers and ranchers in the Skeena and Buckley valleys would benefit from an expanded local market for their products. Business owners in Prince Rupert, Terrace, Hazelton and Smithers all expected to profit from the anticipated 2,500 new residents. Here indeed was an "economic turning point" for the entire region, boosters enthused.[13]

Plant construction, involving from 600 to 1000 workers, shifted into high gear by the summer of 1950. But some negative consequences accompanied Prince Rupert's boom. The influx of workers created a severe housing crisis, overtaxed medical facilities and generated unemployment as the labour supply exceeded the number of construction jobs. Growing pains could be endured, however, given the long-term benefits to be reaped. And nothing bred more confidence than the firm's management licence, which promised a perpetual supply of raw material. "There will never be any danger of the plant closing for lack of timber because the Company has leases on reserves to last 200 years," Arthur Downs proclaimed.[14]

Almost from the outset, however, Colcel's logging and forestry program experienced difficulties. The plant, British Columbia's third largest behind Ocean Falls and Powell River, would require 250,000 feet of logs per day. Meeting that target involved an extensive road construction program, use of the Skeena River for log driving, purchases from contractors, and towing from coastal holdings. Colcel began by purchasing logs cut by Terrace contractors along the Kitsumgallum River, perhaps in an effort to ease concerns about their place in the new timber economy. By the summer of 1950, some 10 million feet had been decked along the Kitsumgallum in preparation for driving down to the Skeena, with plans to then complete the drive down to the mouth of that river.[15]

The use of rivers for log transportation, an established practice in eastern Canada and the southern interior of British Columbia, had never been tried on this sort of scale on the Skeena. Colcel brought out Bert Tremblay, an experienced Quebec woods manager, to conduct its Skeena operation. After a period of study Tremblay decided to proceed with the drive in August, but the river dropped unexpectedly, stranding perhaps 30 per cent of the logs on the gravel bars and banks of the Skeena. Marc Gormely described the attempt as a "complete failure", and while the company announced that further drives would be carried out, the next year Colcel began using the CNR to carry logs west from Terrace.[16]

Countering this disappointment, Colcel officially opened its Watson Island pulp mill on June 12, 1951. President Harold Blancke praised Premier John Hart, Lands Minister E.T. Kenney and Chief Forester Orchard for their far-sighted efforts to "conserve natural resources and develop the dormant wealth of the province". The firm's management licence placed Colcel and its government partner in the vanguard of the movement to develop the province's resources for the general welfare, and his organization accepted its sustained-yield responsibilities. But behind the public rhetoric of partnership a less harmonious relationship between Colcel management and the Forest Service developed. According to Gormely, "friction between Departmental policies and the policies of the licensee of Forest Management Licence No. 1" had become a problem in his District.[17]

The Colcel plant's opening coincided with the first full year of operations on FML No. 1. In addition to building 22 miles of road north from Terrace into its Kitsumgallum block, operations on Khutzeymateen Inlet got underway, and engineers began surveying roads and cruising timber in other blocks. Production reached over seven million feet in 1952, with district staff reporting better forestry performance by the company. Cut blocks had been laid out to ensure reservation of seed sources and firebreaks, and the conduct of logging met satisfactory standards. Colcel practiced a highly mechanized form of logging from the start. Fallers and buckers used power saws, a technology that came into widespread use in the immediate postwar years. Caterpillar tractors or high-lead systems accomplished the first stage of log transportation, depending on terrain and timber conditions. By the mid 1950s self-propelled steel spars had begun to replace spar trees on overhead shows, bringing cost savings through the elimination of rigging crews. Mobile cranes loaded the logs onto trucks for transportation to Terrace for transfer to CNR cars, and tugs towed booms from the booming grounds at the mouth of the Skeena to Watson Island.[18]

Logging activities centred on the Terrace area when the firm pulled out of Khutzeymateen Inlet in 1954 after cutting the easily accessible timber there. By 1955 Colcel's 213 loggers in the Terrace district supplied 70

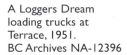

A Loggers Dream loading trucks at Terrace, 1951. BC Archives NA-12396

million feet of timber a year to the Prince Rupert mill, as logging edged northward toward the Nass River. The population of the village and its surroundings reached about 4,000 at this time, with most residents depending on the timber industry for employment. Together, Little, Haugland and Kerr, the Pohle Lumber Company, and Sande Lumber Mills cut roughly 18 million feet a year and employed about 200 men. But Colcel's tight control over the resource in the Skeena River valley forced area sawmillers into fierce competition for timber supplies. By the end of the decade Little, Haugland and Kerr had sold out to a Chicago company that closed the sawmill to focus on pole operations near the Kitsumkalum River.[19]

Colcel's arrival also failed to provide a more competitive market for independent loggers on the coast. The Powell River, Ocean Falls and Prince Rupert pulp mills were the only log buyers, and the last operation proved reluctant to accept cedar or spruce by the early 1950s. Pacific Mills boomed its choice sawlogs for rafting to the Vancouver market, but that did not prove economically feasible from Port Edward. The elimination of Colcel purchases prompted an attempt by some of the larger independents around Prince Rupert to crack the Vancouver market, but low-grade timber and towing costs made this a break-even proposition at best. "The industry on the coast is in effect controlled by the pulp producing companies who directly control a large enough portion of their log requirements that their log purchases can be on a take it or leave it basis," Gormely reported in 1955.[20]

Only an infusion of sawmills would create more competition for coastal logging operators who lived a precarious existence as contractors for Pacific Mills and the Powell River Company. But sawmill production

Transferring logs from trucks to railcars for shipment to a Colcel mill, 1951.
BC Archives NA-13027

remained "of only very minor importance" on the coast, and the prospect of a more diversified forest economy dimmed as corporate concentration increased during the 1950s. The construction phase of the Alcan aluminium project at Kitimat sparked a brief flurry of activity for a new sawmill there in the early part of the decade, along with contract logging around the area to be flooded in developing the massive hydroelectric facilities needed to power the new smelter. The project consumed a sizeable portion of Tweedsmuir Park, brutally displaced the Cheslatta people from their ancestral lands and created a new coastal community of some 9,000 residents. And its completion in 1954 left north-coast loggers once again at the complete mercy of the three large pulp manufacturers who set regional log prices. District staff predicted: "Until, if ever, we have sawmills or wood producing plants in the north, the price will undoubtedly continue to remain stationary."[21]

But if log prices along the north coast remained static, the same could not be said for the rate of cut. Log production in the Prince Rupert Forest District rose to over 500 million feet in 1953, a new high. The district's coastal region was "just coming into its own", Percy Young remarked, thanks to the opening up of new operations as far north as Stewart, increased production on the Queen Charlottes, and higher cuts on the part of the pulp companies. The demand for Crown timber reached unprecedented levels in the interior as well by the end of 1954, as the Forest Service began regulating the cut based on the newly established Public Sustained Yield Unit (PSYU) in the Smithers and Babine Lake areas. But mounting pressure to lock up the resource along the coast drove a frantic process of industry integration and concentration. In 1950, after receiving FML No. 2, Pacific Mills had joined with the Canadian Western Lumber Company to build a newsprint mill near Campbell River on Vancouver Island. The new Elk Falls plant entered production in 1952, and the following year Pacific Mills' parent company Crown Zellerbach acquired Canadian Western, along with its extensive timber holdings and huge Fraser Mills sawmill, merging the new holdings with Elk Falls and Ocean Falls into the new Crown Zellerbach Canada. Crown Zellerbach established a far-flung transportation system to knit its various coastal enterprises together. New self-dumping log barges replaced Davis rafts in carrying logs from the Queen Charlottes to Ocean Falls and Fraser Mills, while other barges carried sawmill waste to the pulp mills.[22]

The creation of Crown Zellerbach Canada was part of a massive restructuring of the coastal forest industry that saw new integrated corporations jockey for control over a dwindling resource. H.R. MacMillan merged with Bloedel, Stewart and Welch in 1951, creating MacMillan Bloedel, a huge organization that operated pulp mills at Port Alberni and Nanaimo. The Koerner brothers shared ownership of Alaska Pine and Cellulose with the Abitibi Paper Company until 1954, when Rayonier took control to become, along with Crown Zellerbach, one of the two largest American-based companies in the province's forest industry. Canadian Forest Products and B.C. Forest Products joined those firms at the head of an increasingly consolidated industry, and then in 1960 MacMillan Bloedel merged with the Powell River Company to create Canada's largest and most highly integrated forest products company.[23]

The corporate reshuffling of the forest industry set the stage for heavy competition over north-coast timber supplies just as mounting concern over monopolization prompted W.A.C. Bennett's Social Credit government to announce a second inquiry by Sloan. By the early 1950s over 20 FMLs had been awarded, another 43 applications had been advertised, and 37 more were at a preliminary stage. Colcel's licence remained the only one in the Prince Rupert Forest District, but 9 other applications had been received, and the submission of another in 1955 would "almost complete the alienation of the readily accessible coast timber north of Rivers Inlet," if all were granted.[24]

By this time opposition from small operators around the province to the monopolistic nature of the management-licence program had become a roar, and rumours of ministerial impropriety in the award of tenures had reached the legislature itself. "Money has talked," Liberal MLA Gordon Gibson charged, referring to Social Credit Minister of Lands Charles Sommers' handling of B.C. Forest Products' application for a Vancouver Island FML. Sommers would later be convicted for accepting a bribe, but in the interim the second Sloan Commission got underway in a highly contentious atmosphere.[25]

FML holders such as Crown Zellerbach, Alaska Pine and MacMillan Bloedel took pains to emphasize the burdens, risks and responsibilities associated with the tenures. Crown Zellerbach described the policy as "an honest effort to team up industry and government" in sustained-yield management. Alaska Pine's Walter Koerner characterized the FML agreement as a unique plan for "public-private cooperation" in which industry "assumes serious obligations as a trustee for the public". H.R. MacMillan, who had harboured doubts about the tenure, now countered the perception that it represented a "give-away policy" by the government. Forest

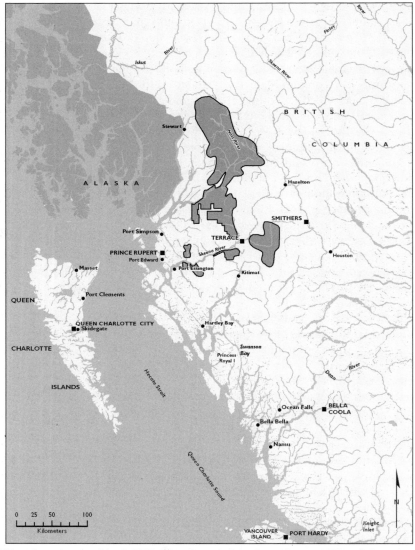

Tree Farm Licence No. 1, 1959 (after *Report of the Forest Service*, 1959, p. 24) secured enormous areas of Crown forests for its holder, Colcel.

A Crown Zellerback log barge at Moresby Island, 1962.
BC Archives NA-21393

Service approval of cutting plans constrained his firm's freedom, and the responsibility for establishing a new forest on lands denuded by logging, fire, insect attack and disease imposed a serious financial burden.[26]

Sloan presented his report in September 1957, proposing only a couple of substantial revisions to the FML tenure. The charge of monopoly control ignored the tenure's contribution to "the permanent, sustained, and increasing production of manufactured commodities," he declared. True, the government had failed to develop clear criteria in awarding the tenures, but though not free of defect the sustained-yield policy met "the pragmatic test – it works," he concluded. Sloan recommended that pending applications requiring only working plan approval be accepted, but that the awards be limited to a renewable duration of 21 years to ensure adequate government control. Amendments adopted in 1958 incorporated that notion, and awards continued under Sommers' successor Ray Williston. Another change required the holder of a tenure, now known as a Tree Farm Licence (TFL), to give contractors the opportunity to harvest 50 per cent of the allowable cut attributed to Crown lands.[27]

Sloan's endorsement of the TFL paved the way for the tenure system's expansion along the north coast. Approval of Alaska Pine's (Rayonier) TFL on South Moresby Island placed that firm in control of nearly 300,000 acres of timber there, despite the 16 protests filed. "The trickle of operators into this District from the Vancouver District ...is now becoming more or less a flood," Percy Young reported in 1956.[28]

The locking up of northwestern resources fostered a good deal of resentment from both aboriginal and non-aboriginal people. One source of discontent involved the disruption of traditional trapping activities by

Kitwanga, 1947.
BC Archives I-28231

logging, which destroyed the habitat of fur-bearing animals. Harold Sinclair, a Gitxsan trapper at Kitwanga, complained about pole cutting on his people's trapping grounds in 1948. Such lands belonged to his ancestors, Sinclair argued, and had provided his own family with a living for the past 20 years. Gormely replied that Sinclair's trapping licence conveyed no title to the land, but that he would advise pole contractors to keep his trails clear of felled timber. "I was born on my property as an Indian," Sinclair responded, and if such a great need existed for the timber he would gladly cut it himself rather than have someone else derive the benefit from ruining his trapping grounds.[29]

Sinclair's protest hinted at growing tension in the relationship between the Gitxsan people in the Kitwanga-Kitwancool area and an unsympathetic Forest Service. Aboriginal people along the Skeena River had been cutting poles for the Hanson Lumber and Timber Company for some time, a few taking out timber sales with funds advanced by Hanson. In the mid 1940s, apparently disappointed with the interruption of pole cutting by salmon fishing in the spring and summer, Hanson reportedly stopped funding them, preferring to take out company sales and hire them as needed. In 1950 the Hazelton forest ranger reported that only one aboriginal between Kitwanga and Kitwancool had been able to put up the necessary deposit to acquire a sale, a failing he attributed to their tendency to squander earnings from fishing. That year Sinclair and the Kitwancool band petitioned the government for a forest reserve, but a Hazelton ranger ridiculed the request in replying to Marc Gormely's request for information. "These Indians come back from the coast in the fall and loaf and drink home brew until about Christmas time, when all funds, credit and everything is exhausted, after which they might do a little work if it is offered to them on their doorstep," he said, calling Sinclair "shiftless, a schemer" whose influence did not extend beyond the illiterate to the "more educated or better Indians".[30]

Minister of Lands E.T. Kenney turned down the Kitwancool request for a timber supply, citing Crown forest regulations that mandated public competition for timber sales. Making no reference to the fact that the government's TFL policy contradicted this commitment to competition, Kenney did agree to have Gormely contact Sinclair to discuss possible solutions. Gormely did so, telling Sinclair that the Forest Service would be willing to consider long-term timber sales to First Nations in the Kitwancool valley. But the offer may not have been entirely genuine, as Gormely was confident that the issue would expire on its own accord. "I anticipate that little will come of this demand from the Indians since they rarely have sufficient funds to acquire timber," he informed Orchard. Nor would Hanson or other major contractors put up the money unless the company's name appeared on timber-sale contract.[31]

B.C. Forest Service ranger station at Hazelton, 1949.
BC Archives NA-09841

The ranger, nevertheless, did meet with Kitwancool band representatives and the Hazelton Indian Superintendent. When the Kitwancool expressed a desire to avoid the requirement that timber-sale applications be advertised in local newspapers, he refused to issue assurances. Agency policy insisted on advertising where any chance of competition existed, and all purchasers of Crown timber must pay stumpage fees, he explained. Area First Nations had long been aware of this policy, the ranger claimed, one they now sought to circumvent by obtaining a reserve directly from the Minister. "Indians everywhere claim all the areas around their villages as their own," he concluded. "We should treat them the same way as we do anyone else."[32]

Kenney was more diplomatic when the federal Department of Indian Affairs Superintendent endorsed the Kitwancool band's request. His department had no authority to reserve timber without competition to any interest except under the TFL tenure, he asserted. Moreover, members of the band in question had no past difficulty in securing timber sales when they so desired. The band could apply for a TFL, Kenney suggested, but it seemed unlikely to possess the qualifications necessary to undertake the required forest management responsibilities. Perhaps the Department of Indian Affairs would like to discuss accepting such responsibilities and apply for a TFL on their behalf, he suggested, in an effort to pass the ball back into the federal court.[33]

Sinclair continued to press the issue in early 1951 as a Skeena District representative of the Native Brotherhood of British Columbia, complaining of the Hanson company's ability to control timber supplies by bidding up the price on advertised sales. Kenney sent the standard response: First Nations were at "perfect liberty" to apply for Crown timber, but would receive no special privileges. A petition bearing the names of 16 Kitwancool contractors followed at the end of 1952, expressing a desire to avoid taking "drastic action" over an alleged threat by Hanson management to use any means to obstruct aboriginal access to timber sales.[34]

Another ranger, J. Mould, attributed the petition to the recent arrival of Hazelton Forest Products, which followed Hanson's practice of acquiring timber sales and turning pole operations over to aboriginal contractors. This process might begin with an aboriginal logger locating the timber, making an application and accompanying the ranger on the cruise. But advertisement of the sale frequently saw the companies step in to "bid the prices up high", assume control of the timber, and finance the sale from the profit margin in buying the poles from the original First Nations applicant, now reduced to contractor status. "The natives are thus caught in the middle," Mould noted, "since it is they who operate and pay the stumpage." Moreover, most were illiterate, and lacked sufficient knowledge of timber-sale procedures to compete. District Forester Percy Young informed Victoria that the "independent" Kitwancool "still tend to consider the region ... as their personal property." Young wanted a response from Orchard, but headquarters staff could only suggest another meeting to further explain provincial policy.[35]

The matter seems to have lain dormant until the end of 1954, when Skeena MLA Frank Howard spoke up in response to Hazelton Forest Products' practice of bidding up prices on cedar pole sales beyond the financial resources of original applicants. That system had given the firm "quite a control on the logging industry in the Hazelton-Kitwanga area," Howard observed, confining most small operators to a subservient position. The company now apparently insisted that even the smallest sales in isolated areas be disposed of by public competition. Howard and a group called the Kitwancool Timber Sale Contractors requested that the Forest Service withhold small sales from the auction process, but District Forester Young replied that regulations tied his agency's hands.[36]

The Forest Service received yet another, more comprehensive Kitwancool petition in late 1955. This statement urged that area First Nations be given priority in timber allocation on the grounds that the land-title controversy with the province remained unsettled. No further sales should be made until resolution of the issue, and the Colcel TFL should be cancelled until a satisfactory settlement of the land question had been reached. The firm's operations were "wasting our timbers and young trees and will also destroy our traplines and fur industry," insisted the Kitwancool.[37]

No record of the Forest Service's response exists in the file covering this dispute as it unfolded between 1948 and the end of 1955. But Percy Young considered it regrettable that the Forest Service could not provide First Nations in the Smithers and Babine PSYUs with protection against the companies that controlled the resource. "It is unfortunate that we must deprive them of their opportunity to operate on their own in the vicinity of their villages," he concluded. The same story played out along the north coast as policies geared to promoting large-scale industrial capitalism severed First Nations from traditional seasonal cycles of resource use and deprived them of entrepreneurial access to the forest.[38]

Georgetown, 1962.
BC Archives H-05594

Trapping went into a "steep economic decline" in the 1950s, and hand-logging opportunities dwindled along with the supply of timber accessible from shoreline and the trend toward corporate control of the resource. The Tsimshian had adapted by applying fishing boats to a seasonal cycle of fishing, gathering and handlogging for a time, but curtailment of access to resources by rigid property laws and the increasingly capital-intensive nature of commercial fishing meant a mounting reliance on wage labour. Colcel provided logging and mill work for the Tsimshian at Port Essington and Port Simpson during the 1950s and 1960s, and when the former community burned down in the late 1950s some moved to Port Edward. The Georgetown mill and Brown's mill on the Ecstall River also relied heavily on Tsimshian labour during this period, and the logging of reserve timber drew some into the forest sector.[39]

But throughout this period, Forest Service policy held to the provincial Crown's unwavering position that aboriginal title had been extinguished, and First Nations held no special claim to resources. In 1961, the Kitwancool responded by blocking a Forest Service road-construction project into the Cranberry River area, and two years later they protested a Little, Haugland and Kerr operation in the same region. The Forest Service announced that aboriginal land claims were a federal matter, a presumption that came under increasing pressure during this period as Frank Calder revived the Nisga'a peoples' claim to title by uniting the communities of Kincolith, Greenville, Canyon City and Aiyansh under a new tribal council. Calder, the first aboriginal person elected to a Canadian legislature in 1949 as a Cooperative Commonwealth Federation MLA for the Atlin riding, became a chief in 1958 and re-energized the Nisga'a claim to both the land and resources of their traditional territory.[40]

It seems certain that Colcel's TFL No. 1 provided some of the motivation for Nisga'a militancy. Colcel hired about 300 Nisga'a on its operations, but over one-third of its tenure occupied Nisga'a land. In 1958, having exhausted its timber around Terrace, Colcel undertook a major road-construction program north toward its Nass River holdings. By this time the Prince Rupert mill employed 500 workers, and in meeting its ever-increasing fibre needs the firm adopted an aggressive style of clearcutting that sometimes disturbed Forest Service officials. Between 1952 and the end of 1956 the firm had cut almost 62 million feet of timber, within its allowable annual cut (AAC) for the period, but the 1956 cut had exceeded the AAC for the first time. Colcel had also proven somewhat loose in its interpretation of cutting permits, which governed annual operations on the licence. When penalized for a trespassing infraction on the Kalum block, company managers argued that the cutting plan was a general scheme only, not to be rigidly adhered to but open to change without the district forester's approval. "It is believed that the licensee is aware of the correct interpretation of the cutting permit condition," Percy Young remarked.[41]

By the end of 1956 Colcel had constructed 146 miles of road, and its main route north, the Nass Road, had reached Lava Lake. Progress continued for the next two years, and by the end of 1958 Colcel's road network had crossed the lava beds at the north end of the lake, branching west to the Andegulay block and east to the Aiyansh block of TFL No. 1. Colcel even managed to enlarge its tenure during this period. After dropping some coastal blocks from the licence in 1959 and enlarging its interior holdings, the productive area of the licence totalled 825,212 acres. Colcel reached the Nass in 1959, completing its 70-mile link from its Terrace logging headquarters to the Nisga'a village of Aiyansh. The firm initially used tractors to yard logs cut from nearby stands to the river, taking advantage of level ground, but turned increasingly to mobile spar trees on rougher ground. Log drives on the Nass involved the use of tugs to assist transportation from the booming grounds to the mouth of the river for collection into rafts for the 72-mile tow to the mill. When driving the lower Nass proved difficult Colcel developed an 11-mile system of side channels to avoid the sandbars and jams that obstructed passage. Boom deflectors on the main river directed logs past sandbars, while two-man boats patrolled the river to break up minor jams before major obstructions developed. By 1964 the firm had invested a reported $6 million in building 600 miles of main and access roads, introduced a range of sophisticated logging equipment, and had altered a major salmon ecosystem in meeting its fibre needs.[42]

Colcel's arrival on the Nass produced enormous changes for the Nisga'a, providing employment but destroying many traplines. As representative Rod Robinson explained to the Senate-House of Commons Committee on Indian Affairs in 1960, the Nisga'a welcomed industrial expansion but demanded compensation for the damage inflicted on their territory. The Nisga'a launched their land-claim action against British Columbia in 1967, but in the interim Colcel continued carrying out what Daniel Raunet calls a "scorched earth policy" in the Nass watershed.[43]

Forest Service regeneration studies on TFL No. 1 produced less alarming findings, but did give some cause for concern. A 1957 survey of areas

Nass River
log drive, 1973.
BC Archives NA-28892

logged between 1950 and 1952 found that the growth of heavy brush on some areas would make planting impractical. Colcel began a very small artificial reforestation program that year, planting cottonwood on 30 acres. The company followed up with 98 acres in 1958, and established a small nursery the following year. Natural regeneration of cutovers in spruce-balsam stands appeared to be "progressing satisfactorily", District Forester J.R. Johnston reported in 1960. But even where reservation of seed sources was adequate, brush quickly overtook the most productive river-bottom lands, and the firm ignored site preparation needed to prepare non-stocked areas for planting. A 1960 survey of 11,500 acres cut over the 1950-1958 period found some 8,000 naturally restocked, almost 3,000 without new growth, and only 234 acres planted.[44]

Reports generated during the early 1960s indicated a reasonable standard of natural regeneration on cutover hemlock-balsam stands, but serious understocking of river-bottom sites continued to pose a problem due to the rapid establishment of deciduous brush. Within Forest Service ranks, then, Colcel's forestry practice seemed acceptable if not outstanding. By the end of 1961 the firm had planted a total of 410,000 trees on 1,040 acres on TFL No. 1, clearcutting 3,840 acres that year alone. Clearly, a good deal of faith had been placed in the land's capacity to produce a second crop of commercial timber despite the aggressive nature of Colcel's clearcutting.[45]

First Nations were not alone in expressing resentment both at the control the multinationals had achieved over the resource, and the conduct of their operations. After having his timber-sale application in the fully committed Smithers PSYU disallowed a Buckley valley rancher complained that he was "absolutely unable to compete with other timber interests

Clearcutting operations at Kitsumkalum Lake, 1973.
BC Archives NA-28897

here." The area's ranchers and farmers needed seasonal woods work to supplement their incomes, he explained, and their use of horses and small tractors was "selective logging in the truest sense", preferable to the large operators' bulldozers which destroyed "young timber by the acre". The large interests had succeeded in penetrating the northwest, seizing millions of feet of timber, but contributed "nothing to this country", W.C. Gardiner concluded.[46]

Forest Service reports of changing industry structure in the interior during the late 1950s lent at least some support to Gardiner's observations. Large-scale operations were taking over production around Terrace, Smithers, Telkwa, Houston and Babine Lake. Tractors assembled logs for trucking to larger stationary sawmills equipped with efficient gang saws. By 1958 truck hauls had reached 20 miles in some areas, as the portable bush mill gave way to larger, more heavily capitalized plants. At the top of the structure stood the few planer-mill operators who owned sawmills, purchased logs from independents, and thus controlled the market for both logs and lumber. By 1960 all interior operators were aware of the dwindling supply of easily accessible timber, triggering more intense competition for holdings. The introduction of the "licensee priority system" in the PSYUs, which gave established operators preferential access to the allowable cut of these units, contributed to competition among non-TFL holders. Such "quotas" quickly became saleable commodities in a market dominated by the larger companies, who consolidated their position by purchasing the timber sale licences of smaller competitors.[47]

The quota system worked its logic throughout the province's forestlands, a process that helped TFL holders extend their control over Crown timber beyond the boundaries of their already massive holdings. And the scramble for TFL tenures picked up as the timber supply along the lower coast dwindled year after year. B.C. Forest Products, the Allison Logging Company, Crown Zellerbach and MacMillan Bloedel, among others, pressed applications along the north coast. But could the region withstand the flood of operators? Percy Young found it difficult to believe that the Prince Rupert District's coastal supplies were sufficiently abundant to take in all the eager applicants. "A large portion of the best timber is presently covered by Management Licence reserves," he reported, "while the remaining timber is in small, narrow valleys or along the shoreline." Young's reservations notwithstanding, the process of consolidation continued unabated under W.A.C. Bennett's Social Credit government. After a brief hiatus during Sloan's second Royal Commission, the majors engaged in a cut-throat competition to lock up the best remaining parcels along the north coast. Alaska Pine emerged victorious, winning the province's 25th TFL in 1958, encompassing several blocks of timber in the Vancouver and Prince Rupert districts. MacMillan Bloedel succeeded in securing TFL No. 39 in 1962, involving over a million acres in tracts on northern Vancouver Island, the mainland coast and the Queen Charlotte Islands. The firm announced its Intensive Forestry Program at the same time, committing $5 million to increase the productivity of its forestlands.[48]

But the vast bulk of the timber from these tenures, along with that from the old licences and new timber sales, was destined for southern plants. Percy Young had questioned the policy of alienating north-coast timber to supply south-coast mills in 1955, as E.C. Manning had decades earlier, but the government did little or nothing to halt the trend. The widespread adoption of self-dumping barges to move logs efficiently from as far north as the Queen Charlottes to Vancouver encouraged the flow of wealth out of the region. Only construction of the large Prince Rupert Sawmills on the city's waterfront in 1960, drawing logs from a Skeena River timber sale, brightened the outlook for fuller local utilization of the resource and a more competitive log market. The firm negotiated a deal to sell chips to Colcel, found Japanese customers for its lumber products, but struggled to generate the investment capital needed to develop the timber sale.[49]

Nowhere did the political economy of postwar forestry manifest itself with more clarity than in the Bella Coola River valley. Crown Zellerbach ruled in that region by virtue of its old tenures and more recent timber sale acquisitions, drawing timber from the valley to the Ocean Falls and Elk

The *Forest Prince*, a self-dumping log barge, 1962.
BC Archives NA-21532

Falls paper plants and the Fraser Mills sawmill. Indeed, back in 1947 Cliff Kopas had noted an oddity that captured Bella Coola's hinterland status perfectly. There, despite an immense forest and busy logging activity, a local lumber shortage retarded community progress. Bella Coola's main locally-based enterprise, the Northern Cooperative Mill and Timber Association, responded to dwindling membership by forming a limited company in 1948. Fourteen shareholders controlled the new Northcop Logging Company, building a diesel-powered sawmill and planer in 1952. Northcop also logged under contract to Crown Zellerbach, but when that company initiated its own Bella Coola operations in 1955 problems in acquiring timber caused a decline in sawmill employment.[50]

That development may have contributed to a mounting concern over the community's place in the evolving industrial structure. An unsustainable rate of logging and waste of the resource topped the list of Bella Coola's grievances. Fear of forest depletion prompted the local Board of Trade to approach Crown Zellerbach, "the big moguls of the Bella Coola woods", requesting the construction of a chipper plant to utilize wood left behind after logging. Processing of slash would generate employment, reduce the fire hazard and "extend the forest wealth of the District". Crown Zellerbach declined, offering instead to pay $12 to $15 per cord for small logs delivered to the waterfront, but the board countered that 70 per cent of the wood left on the ground did not reach the minimum 26-foot length the firm would accept.[51]

At the same time, the board asked the Forest Service to survey the valley's resources as a first step toward establishing a PSYU that would place cutting on a sustained-yield basis. "Most of the talk surrounding 'perpetual yield', 'reforestation' and 'full utilization of the forests', etc. etc., Bella Coola considers as mental opiates meant to numb the senses," Kopas observed. The community envisioned a future "not as ... the nest for some industrial octopus," he stressed, but as a "little Switzerland, where free people live in

Logging trucks at the Bella Coola Fair, 1950s.
BC Archives D-03100

the midst of mountain beauty, and with their manifold small industries are free, happy and industrious."[52]

The difficulty Northcop and the region's five or six other small operators experienced in gaining access to timber reflected how sharply reality failed to square with Kopas's vision. In 1956 Northcop's Stenner Saugstad asked the Forest Service to make sufficient timber available to operate the mill for 10 to 15 years while selling the best logs at Vancouver. Recently Crown Zellerbach had made only poorer logs available to Northcop, but a certain tract on the south side of the Bella Coola valley would permit the operation to meet local needs and ship lumber to the Vancouver market. Northcop had the necessary equipment and manpower, and could "better utilize this timber than other parties," Saugstad advised.[53]

But Northcop's plea for an assured timber supply met with a lukewarm response, dictated by the constraints of a forest policy that seemed to make the link between resources and local economic development only in the case of large enterprise. Any applications by Saugstad's organization would have to go through the auction process, Young replied. Moreover, Crown Zellerbach had recently included the area in question in a TFL proposal. Since processing of those applications was on hold pending completion of the second Sloan Commission, and granting Northcop's request might affect the sustained-yield capacity of the area, the Forest Service refused to make the timber available.[54]

Young did agree to provide for Northcop's immediate requirements, up to two years of operation, but Saugstad found this offer unattractive. The company's inability to buy suitable timber at a reasonable price made profitable operation impossible, and closure inevitable. "We feel that after being established here since '39, and are natives of the valley, that we

Hauling logs with a rubber-tired skidder, 1968.
BC Archives NA-24202

should not be forced to close down ... due to large and new companies establishing their operations here," he emphasized in renewing his application. But the Forest Service again declined, citing potential effects on the allowable cut of the proposed TFL.[55]

Frustrated in achieving its long-term objectives, Northcop went ahead with a number of timber-sale applications in 1957. The Forest Service awarded the company one "direct" sale outside the usual auction process, but by year's end Saugstad revived his mission, this time going directly to Minister of Lands and Forests Ray Williston. Northcop operated the only sawmill, provided one of the largest payrolls and supplied a good deal of the lumber required in the Bella Coola valley, he told the minister. Yet the company now found itself in a precarious timber supply position, owing to recently arrived operators "interested mainly in logging the available timber and moving on with no thought to the future". Citing discrimination, Saugstad went on to assert that four of the five timber-sale applications made over the past year had been refused, two of them subsequently granted in part to other operators. Given Northcop's "balanced operation", modest timber needs and willingness to reduce production to fit into a sustained-yield program, would the Forest Service afford the business some consideration?[56]

Williston requested a full report on Saugstad's inquiry, but offered no concessions after reviewing the case. The sales had been declined because of immature timber values, and he denied any discrimination against Northcop. Percy Young thought that the valley was best suited for management as a PSYU, however, given the number of parties involved in the

Rayonier steel spar, 1962.
BC Archives NA-21402

area. In the end, Crown Zellerbach did not gain approval of its TFL application, but the number of contending interests around Bella Coola made
for a chaotic situation by 1960. "The timber available to operators established in the Bella Coola Valley is becoming extremely scarce," Young's
successor J.R. Johnston noted. "Competition for the available timber has
grown to a point where some operators have already been forced out to the
channels ... and much greater migration of operators to the channels is
expected." Even after the valley's inclusion in the Dean PSYU, established
in 1960, the annual cut exceeded sustained-yield limits. Northcop managed
to survive until 1987, cutting for local and export markets, but the Bella
Coola valley remained a stronghold of corporate logging to supply mills
located elsewhere.[57]

    While the Forest Service managed affairs in the contentious Bella
Coola area, the agency extended its PSYU system in the northwest. By
1960 the Dean, Kitimat, Hecate and Rivers Inlet units had come into existence along the coast, in addition to several interior jurisdictions. Annual
cuts in most remained within sustained-yield boundaries for the time being,
but overcutting on the Terrace unit posed a serious problem, and the award
of large sales on the coast made future production increases certain. Prince
Rupert Forest District staff scaled 777,558,000 feet of timber in 1960, a

record that eclipsed the 1959 cut by 44 per cent. The cut went up again in 1961, prompting J.R. Johnston to plead for personnel to cope with the ever-increasing administrative burden. "To adequately supervise the present activity we should have more ranger districts and staff and more professional foresters," he warned. "Unless an increase in establishment is realized soon the situation will be critical."[58]

The Forest District's scale of responsibilities kept rising in the early 1960s, however, as the agency's Engineering Division continued pushing roads into virgin forests and construction of the Stewart-Cassiar road brought new interest in northern timber. Operators applied for nine large sales in the Unuk, Stikine and Iskut river drainages in 1962. Total District production that year reached 905,700,000 feet, a year of intense logging by TFL holders and timber-sale operators who clearcut over 27,000 acres. A loose form of selective logging in the interior brought the total cutover area to almost 42,000 acres, but mechanization limited the growth in logging employment. Output had almost doubled since 1955 when 2,784 loggers turned out 553,559,000 feet, but just 2,860 men produced the 1962 cut of over 900,000,000 feet.[59]

Along the coast, operators employed a mix of old and new techniques to lower labour costs in a region that presented significant topographical challenges. A-framing remained important on the dwindling number of tidewater shows, while inland the portable steel spar dominated high-lead operations. Tractors and rubber-tired skidders might be used in conjunction with this equipment to yard logs short distances from spars to the water. Trucks handled longer hauls, their use increasing along with the distance separating logging sites from the inlets and bays that served as booming grounds. "Extensive use is made of mobile steel spars, track-side settings, mobile loaders, and huge logging trucks on the larger operations to cut costs and increase utilization," District Forester N.A. McRae observed in a 1962 summary of north-coast logging.[60]

The drive to reduce costs and increase efficiency also promoted further mechanization of interior logging, which remained largely a winter affair. Notable in this trend was the disappearance of the tie hack, "killed off by old age and prosperity," Young remarked in 1955. Only 13,000 hewn ties were produced in the Prince Rupert District in 1961, the last year the CNR issued contracts. As the skills of the tie hack faded, others more geared to machine operation continued to expand. Truck logging also became more prevalent in the interior, due to the elimination of portable bush mills in favour of larger plants located in the communities of Terrace, Hazelton and Houston. Crawler tractors, sometimes equipped with arches, still performed the first stage of log transportation on many operations, but proved uneconomical when skidding distances exceeded a mile or so. Interior loggers then developed their own technology, installing winches and arches on trucks so that the lead end of a bundle of logs could be elevated for hauling longer distances to landings for loading onto trucks bound for the mill.[61]

Patch logging on Graham Island, 1962.
BC Archives NA-21387

Mechanized logging on the coast made clearcutting the standard practice, demanding that the Forest Service regulate the reservation of seed
sources. Typically this means limiting the size of clearcuts and leaving
blocks of timber for a period of years to promote natural reforestation. But
by the end of the 1950s firms were rebelling against such "patch-logging"
requirements, advocating more extensive clearcuts and planting to reduce
road-building and logging costs. "Permanent seed blocks are often a source
of disagreement between the District staff and the larger companies," J.R.
Johnston reported in 1959, and the lack of definite silvicultural data
ensured that the conflict would persist. But in Johnston's view regulation of
logging plans provided "one means of maintaining some control over the
resource so that natural restocking is given a chance."[62]

The fact that the Forest Service provided seedlings from its south-coast
nurseries increased the appeal of plantation forestry in British Columbia,
and the agency soon felt pressure from corporate foresters who cited blow-
down along the edges of cutblocks, the uncertainty of natural seeding, and
cost factors in advocating progressive clearcutting. Prince Rupert District
Forester J.R. Johnston asked Chief Finlay McKinnon for a clarification and
uniform application of agency policy in late 1959. District staff felt strong-
ly that fire breaks and seed sources were essential to protection and reforestation. Alaska Pine and Colcel foresters "unequivocally opposed" this
opinion, he related, and appeared "to be quite happy to have literally
thousands of acres of continuous slash." Victoria officials responded that
promises to plant should be accepted only if accompanied by specific plans
committing the firm to proceed quickly after logging.[63]

The Forest Service's example of poor logging, at Ootsa Lake, 1971.
BC Archives NA-26582

Fulfilment of such obligations would have been difficult at the time, however, as Colcel had only just opened a small nursery, and Alaska Pine had yet to undertake any regeneration activities on TFLs No. 24 and No. 25. The Forest Service itself had nothing to boast about in this regard, not establishing its Telkwa nursery to meet District seedling needs until the late 1950s. And while the agency insisted that all operators of sizeable timber sales submit logging plans at this time, and subjected TFL working plans and cutting permit applications to scrutiny before approval, the issue of seed source requirements remained problematic. "Strict enforcement of the conditions of 'Coast' guidelines is at times made difficult by a plague of unknown factors," Johnston confessed in 1960. "A good deal of study will be required before we may enforce some aspects of the plans."[64]

Expansion of the Telkwa nursery's capacity brought the number of seed beds to 58 in 1961, and the District hired men for the first time to plant 68,480 seedlings. But a regeneration survey on Alaska Pine's TFL No. 24 at that time showed adequate natural restocking on only 60 per cent of the cutover area. The appointment of a permanent nurseryman at the Telkwa facility in 1962, where annual production reached 45,000 seedlings the next year, promised some progress where cutting practices discouraged the establishment of a new crop. But the Forest Service's 1962 planting effort amounted to only 49,000 seedlings on 94 acres, while Prince Rupert District operators logged almost 42,000 acres that year. Even assuming that natural restocking produced a satisfactory second crop on half of the cutover lands, then, the foundation for a future reforestation crisis was

B.C. Forest Service, Telkwa Nursery, 1952.
BC Archives NA-21455

being laid. Moreover, an increasing number of interior timber sales were "designed to be clearcut in some form or other", either in strips or patches, as mechanized operators took poles, pulp logs and sawlogs from stands east of the Cascade Mountains.[65]

But the short-term outlook seemed bright, one of an economy maturing along sustained-yield lines. About 3,500 earned a livelihood in forestry between Prince Rupert and Smithers in 1959. Fishing and fish processing provided another 3,000 jobs on the coast, while mining and agriculture engaged about 3,800 interior residents. The seasonal nature of fishing and agriculture enabled many to combine these endeavours with forestry, the dominant sector of the economy. Prince Rupert's expansion, the establishment of Kitimat, Terrace's emergence as a logging, lumbering and service centre, and the slower growth of Stewart, Hazelton and other small communities brought the region's population to 47,000 in 1958.[66]

The primary engine of this growth was Colcel, a dynamic force for wealth generation, economic integration and ecological change. A 1963 *British Columbia Lumberman* report on forestry along the Skeena River depicted the company's influence in the most glowing terms. "Probably the most striking impression the Skeena Valley imparts … is that of prosperity," the journal gushed. "Nowhere else in British Columbia have the independent logging and sawmill operators achieved such a level of universal well-being." Colcel, briefly renamed Celgar in the early 1960s, inspired this alleged sense of security. Small operators who had struggled to make a profit cutting stands that provided an inadequate proportion of sawlogs now had a market for smaller, low-grade pulp material and chips. Colcel also provided loggers with the opportunity to "round out production" by contracting to the firm, extending the work-year.[67]

Terrace provided the best apparent example of a mutually beneficial relationship between the giant, local producers and communities, according to this analysis. Colcel's logging operations, the town's seven sawmills, three pole plants and numerous small logging operators provided employment for up to a thousand workers. Skeena Forest Products constructed a new mill there in 1959, buying the timber rights formerly held by Little, Haugland and Kerr. After installing a chipper in 1962 the company began selling its chips to Colcel, which increased the capacity of its pulp mill by 15 per cent to an estimated 160,000 tons that year. Hazelton Sawmills also added a chipper to its plant in 1962, linking its fortunes to Colcel while improving utilization. Independents were "almost unanimous in granting credit to [Colcel] for creating the current stability," the *Lumberman* declared.[68]

Such glowing observations ignored the strenuous battle taking place over timber rights, becoming even more intense as a consequence of new TFLs created to support pulp-and-paper developments at Kitimat and Prince Rupert. Alcan and the Powell River Company had begun negotiating to partner in building a pulp mill at Kitimat in the early 1950s. When that project never got off the ground, putting plans for a TFL on hold, the Forest Service made only short-term timber sales in the Kitimat PSYU. But the TFL application remained on file, and by the early 1960s the Powell River Company had merged with MacMillan Bloedel, which revived the proposal. MacMillan Bloedel held about 50 timber licences in the area to contribute to a TFL. Other contenders in the "fancy manoeuvring for position in the Kitimat Valley" included Crown Zellerbach, which had acquired 41 square miles of old timber licences there in 1946, and now prepared to downsize its Bella Coola operations in favour of a 20-year logging program in the Kitimat area. Prince Rupert Sawmills had vague plans for a sawmill at Clio Bay, while the Skeena Valley Timberman's Association worried that a MacMillan Bloedel TFL pulp mill would deprive its members of access to the valley.[69]

Crown Zellerbach commenced full-scale operations at Kitimat in 1963, moving equipment from Bella Coola to meet its annual production target of 35 million feet destined for Ocean Falls, Elk Falls and the Lower Mainland. By the end of the year the Diashowa Paper Company of Japan had also expressed an interest in building a pulp mill at Kitimat. Representatives visited at Alcan's invitation, the aluminium company anticipating sales of land, water and power if the Japanese went ahead. That coincided with MacMillan Bloedel's application for both a TFL and a Pulpwood Harvesting Area, a new tenure developed by Lands and Forests Minister Ray Williston to utilize small, residual interior timber in paper making, to build an $86 million kraft mill at Kitimat, with a $64 million newsprint facility to follow if markets warranted.[70]

The MacMillan Bloedel application involved an enormous four-million-acre area including the Queen Charlotte, Kitimat, Terrace and

Hazelton PSYUs, and parts of the Hecate, Dean and Ootsa units. But MacMillan Bloedel's plans conflicted with a number of other applications, as the scramble for northwest timber rights heated up. Colcel had proposed to build a new kraft mill alongside its existing sulphite mill at Prince Rupert if granted a second TFL. In addition to requesting more timber on the upper Nass and Skeena rivers adjacent to TFL No. 1, Colcel sought timber in the Hazelton, Babine and Takla PSYUs, including areas south and east of Kitimat that MacMillan Bloedel had claimed. Yet another competitor was the Buckley Valley Pulp and Timber Company, a new entity organized by Houston, Burns Lake and Smithers operators with plans to build a $52 million kraft mill at Houston. That group sought a Pulpwood Harvesting Area covering some 13,000 square miles in the Babine, Smithers, Morice, Burns Lake, Ootsa, Hazelton and Takla PSYUs. MacMillan Bloedel insisted that their project could not go forward without the Ootsa timber, and Crown Zellerbach figured into the equation as well. If granted, both the Colcel and MacMillan Bloedel applications would undermine Crown Zellerbach's ability to buy logs from the Prince Rupert area for the Ocean Falls mill.[71]

Williston set hearings for the spring of 1964 to sort out the overlapping claims. MacMillan Bloedel's J.V Clyne spoke for his firm, saying that the kraft-mill project could not go forward unless Williston awarded all of the timber requested. Timber for expansion into newsprint production was essential, but he could not guarantee the facility. That put Williston in the uncomfortable position of having to award timber for a mill that might never be built, and the hearing ended with no decision. He went on to hold similar proceedings on the Colcel and Buckley Valley applications before travelling to Prince George for hearings related to proposals for a mill at Mackenzie.[72]

Industry submissions reflected the high stakes involved in a game destined to produce both winners and losers. The Skeena Timbermen's Association, its members dependent on timber from the Terrace, Kitimat and Hecate PSYUs, expressed reservations about the proposed MacMillan Bloedel TFL. The organization stopped short of denouncing the application, but only if granting the licence did not "jeopardize any of our present or future cutting needs". Crown Zellerbach responded along similar lines, asking the government to ensure that MacMillan Bloedel's award not deprive it of wood necessary for a proposed $30 million expansion at Ocean Falls. The Truck Loggers' Association opposed Colcel's application outright. Williston expressed confidence that sufficient timber existed to satisfy all the applicants, but hinted that awards would be contingent upon firm plant commitments.[73]

In announcing his decision that August Williston gave each of the contenders a slice of the northwest timber pie, but not exactly in the dimensions requested. "After reviewing all the facts it is my opinion that with some give and take there is, within the entire area under review, sufficient

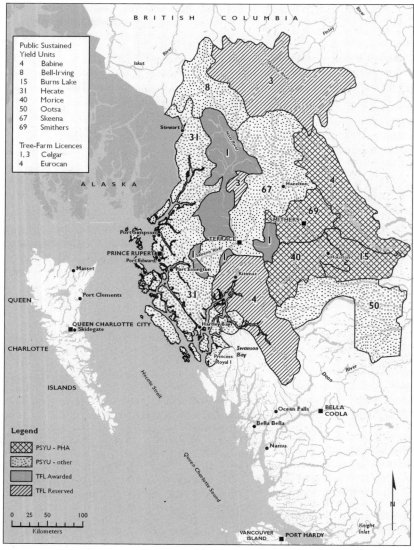

TFLs and PSYUs on the north coast in 1964, based on a *British Columbia Lumberman* report (vol. 48, Sept. 1964, p.17). Celgar was Colcel's name for a brief time in the early 1960s. PHA refers to Pulpwood Harvesting Area.

wood available on a sustained-yield basis to operate all three proposed pulp mills and also to maintain the already existing industry," the minister stated. MacMillan Bloedel got its TFL, obliterating the Kitimat PSYU, but was denied the Ootsa PSYU timber and its Pulpwood Harvesting Area that would have covered the Hazelton, Hecate, Kitimat, Kitwanga, Ootsa and Queen Charlotte PSYUs. Instead of obtaining a second TFL, Colcel's TFL No. 1 would be increased in size to include the Kiteen River watershed and three areas left over after creating the Bell-Irving PSYU. Williston dropped upper Skeena River timber from the Colcel application, adding this timber to the new Skeena PSYU, which encompassed the former Hazelton, Terrace and Kitwanga units. The Buckley Valley group received Pulp Harvesting Area No. 4, superimposed on the Smithers, Babine, Morice and Buckley Valley PSYUs. Although the Hazelton and Ootsa PSYUs were eliminated from the proposal, the firm would be free to compete for pulp material in the area.[74]

The "Solomon's judgement" rendered by Williston left observers stunned. From his perspective, the three new mills would have an assured wood supply to meet immediate production needs. Both Colcel and MacMillan Bloedel could draw about 60 per cent of their annual requirements from the TFLs, and compete for the rest by establishing a quota position in the PSYUs. Those Crown forests would also provide timber for the established sawmill industry. But how would the companies react? He had given them much of what they requested, Pat Carney observed, but rejected a great deal they considered essential. Despite reports that MacMillan Bloedel had lost enthusiasm for the Kitimat project, the firm announced its intention to go ahead in October. Colcel and Buckley Valley had already accepted Williston's terms. So had Prince George Pulp and Paper and Northwood Mills, set to build pulp mills at Prince George, and the Cattermole Timber Company at Mackenzie.[75]

But as the "monumental" year of 1964 wound down, not all signs pointed to a fulfilment of Williston's vision. Buckley Valley lacked both capital and markets, and MacMillan Bloedel apparently came under pressure from the government to lend its financial resources to the venture. By the end of the year Carney characterized Williston's actions over the previous 12 months as "an era of error". Rather than developing a firm policy the minister had "chosen to divvy up available timber among competing users with sometimes bewildering results". At the present rate all of the province's timber would soon be alienated, as Williston sacrificed rational public policy to industry's claims to what remained of a shrinking resource. Citing industry experts, Carney warned that construction of all the proposed mills would destabilize world markets and cause future suffering for dependent communities.[76]

Events at Kitimat briefly eased that concern. In April MacMillan Bloedel announced that it had abandoned plans to go ahead with mill construction because the government had guaranteed only 40 per cent of the required wood supply. MacMillan Bloedel then turned to investment opportunities in Alberta, Europe, the American south and Asia. Williston remained bullish in the face of MacMillan Bloedel's withdrawal and those who questioned the rapid pace of pulp-and-paper development. "I am an optimist," he asserted, and a mill for Kitimat remained near the top of his agenda. News that Prince George entrepreneur Ben Ginter had entered negotiations with Finnish interests to build a mill there provided a welcome indication that the MacMillan Bloedel episode would not interrupt his expansionist designs for the central coast.[77]

Anxious to secure outside sources of fibre to counter depletion of Finland's forest resources, owners of Enso Gutzeit Osakeyhtio, the country's largest pulp manufacturer, heard of Kitimat's availability from a Colcel official. After being introduced to Ginter on an exploratory visit to Kitimat, the Enzo Gutzeit representatives made him president of a new company, Eurocan Pulp and Paper. Ginter took a 15-per-cent interest, Enzo Gutzeit contributed a major share of the capital, and two other Finnish firms jointed the venture. Crown Zellerbach, with major holdings in the area, also came forward to compete for the timber tributary to Kitimat.[78]

The two firms submitted their proposals to Williston during August 1965 TFL hearings at Kitimat, with Eurocan's proving to be most attractive. While Ginter and the Finns promised to have a 580-ton capacity pulp-and-paper facility in production by the end of 1969, the American company only committed to begin construction by 1970. Moreover, Crown Zellerbach, already the operator of two coastal paper plants, would produce only pulp at Kitimat. Williston ruled in favour of Eurocan, citing the earlier start-up date, the additional employment and value generated by paper production, and the eventual sustained-yield management benefits of having Crown Zellerbach continue to log its old tenures while Eurocan harvested the Crown forests.[79]

The forthcoming award of Eurocan's 473,000-acre TFL No. 41 caused some uneasiness among the region's independents. The Skeena Valley Timbermen's Association favoured a pulp mill, provided that members retained the right to harvest sawlogs and cedar poles from the area of the Kitimat PSYU destined for absorption by the TFL. The Truck Loggers' Association filed a similar brief, requesting that half of the TFL logging operations be handled by contractors. Several smaller operators objected to specific parts of the TFL grant on the grounds that their inclusion would eventually decrease the allowable cut of the Skeena and Hecate PSYUs and threaten the existing sawmill industry. But the prospect of adding another major industry to the central coast ensured that Eurocan's fibre needs would be met.[80]

Eurocan signed its TFL contract in December 1966, purchased 800 acres from Alcan for the plant, then encountered various problems that threatened to delay construction. The firm bickered with the municipality of Kitimat and the provincial government over who would pay for the $2-

million road to the site. Ginter promised clearing would begin in 1967, but market conditions took a turn for the worse as overproduction drove down world pulp-and-paper prices temporarily. Eurocan secured an additional $81 million from an American insurance company, but only on the condition that the mercurial Ginter be ousted as president. Design problems plagued the sawmill and pulp-mill projects, forcing the government to grant an extension in the spring of 1968. That set an October 1970 deadline for completion of the projected 760-ton-per-day linerboard and kraft paper plant. At the same time Eurocan received long-term harvesting rights across the Cascade Mountains in the Ootsa PSYU, in addition to its coastal TFL.[81]

While Eurocan went ahead with plant construction and pushed a 30-mile road over the Sandifer Pass from Kemano to West Tahtsa Lake to tap its Ootsa PSYU timber, the region's established powers were also busy by the autumn of 1968. Production by Crown Zellerbach approximated 80 million feet a year, going out over 120 miles of truck road to its Minette Bay log dump for loading onto barges. MacMillan Bloedel had three contractors at work on its licences between the Kitimat River and Lakelse Lake, and anticipated logging 125 million feet annually over the next 10 years. Eurocan had 200 construction workers busy at Kitimat, with expectations of bringing that total to 1,250 by 1970.[82]

The infusion of American capital along with recovery of the world pulp market in 1969 led Eurocan to announce a major expansion as construction proceeded that summer, raising the project's cost to $145 million. Total capacity would be increased 30 per cent, W.A.C. Bennett proclaimed proudly, bringing the daily production of kraft pulp to 915 tons. Two high-speed Finnish paper machines would be installed rather than the single unit originally planned, and the sawmill's annual capacity would double to 150 million feet per year. That brought anticipated employment to 1,400, requiring the construction of 500 new housing units at Kitimat.[83]

Eurocan opened its lumber and chip producing facilities in the spring of 1970, drawing logs from its coastal holdings. The first paper machine went into production in October, and the complex became more fully operational later that year when the second began turning out kraft sack and bag papers. Logging on the relatively flat interior terrain involved tree shears, a recently developed technology that allowed the operator of a crawler tractor to move through the forest clipping down trees. The logs were then forwarded by rubber-tired front-end loaders and skidders to landings for bundling and loading onto trucks for the haul to the dump at Andrews Bay on Ootsa Lake. After crews formed the bundles into rafts, tugs towed them 53 miles through Tahtsa Reach and across Tahtsa Lake at the west end of the water system to a loading station. Contractor Roy Saunders operated 14 trucks from that point, hauling the bundles some 30 miles over the Sandifer Pass Road to tidewater at Kemano Bay. At the end of a 50-mile tow to the head of Kitimat Arm, cranes lifted the bundles from the water and placed them on trucks for the final 3-mile trip to the mill.[84]

Eurocan's pulp mill at Kitimat, 1971.
BC Archives NA-26429

With the enterprise up and running the Finnish consortium bought out Ginter's interest in early 1971. Employment at Kitimat totalled about 800, with up to 200 loggers providing the raw material. Eurocan's use of the latest labour-saving technology limited employment opportunities in the woods. The firm's three tractor-mounted shears were capable of falling a tree every seven seconds, a productivity rate of up to 1,200 logs per shift that more than quadrupled the output of power saw fallers. But the early shears were cumbersome, and early in 1972 Eurocan took another step in the direction of fully mechanized interior logging by introducing one of the early track-mounted feller-bunchers to the province. That technology, consisting of hydraulic shears coupled with a grapple on the end of a boom, permitted operators to cut trees and pile them in rapid fashion. Skidders could pick up the entire bundle and drag it to the landing, achieving "the world's first fully mechanized logging system".[85]

Eurocan adopted the latest technology on its coastal operations as well, employing portable spars with grapples to shed labour on overhead yarding shows. Grapple yarding did away with the choker setters who attached the rigging to logs in the woods and the chasers who released them at the landing, a dramatic advance in mechanization. Now a single machine operator and spotter could produce as many logs as six or seven men on a standard high-lead setting. Installing floodlights on its grapple yarders enabled Eurocan and other firms to commence night-shift logging, maximizing returns on expensive technology.[86]

Despite an impressive level of technological sophistication, operational problems at the paper plant and the complexity of handling materials on its extensive interior woodlands troubled Eurocan. Restricted operation of

Transferring Eurocan's
logs from truck to
water, 1971.
BC Archives NA-26803

the sawmill and pulp facility limited woods operations in the Ootsa PSYU to 60 per cent of capacity until 1973 – this system demanded full production to achieve cost feasibility. By 1974 Eurocan was involved in discussions with the New Democratic Party (NDP) government, negotiating stumpage rate concessions and discussing the prospect of establishing interior sawmills to process timber from that region as a solution to its transportation difficulties. At the same time, Eurocan played a central role in a process that saw the giant companies link fortunes to integrate the mid coast and adjacent interior into a huge lumber-and-fibre-producing basket. Eurocan joined with Weldwood of Canada in building a highly automated sawmill at Houston in the late 1970s, a facility that drew logs from the Ootsa PSYU and routed chips to Kitimat. That followed a joint venture with Weldwood, Canadian Cellulose (formerly Colcel), the NDP government and the Burns Lake Native Development Corporation to construct a sawmill that went into operation at that point in 1975.[87]

By the end of the decade Eurocan, now wholly owned by Enzo Gutzeit, proved attractive to the Ketcham family's West Fraser Timber Company. West Fraser had spent the 1970s buying up a number of small interior mills, cutting lumber and selling the chips to the pulp companies. Unhappy with chip prices, West Fraser partnered with Diashowa in 1979 to build a pulp mill at Quesnel, and negotiations seemed certain to see it assume part ownership of Eurocan. The deal fell through briefly, but merger talks resumed and in 1981 West Fraser, Canada's second largest lumber

Felling with Tree
Shears, 1968.
BC Archives NA-24275

producer, bought a 40 per cent interest in Eurocan. The deal allowed chips from some of the Ketcham's numerous interior mills to be processed at Kitimat, while Eurocan's interior timber went to West Fraser's new Fraser Lake sawmill.[88]

While Eurocan erected a transportation structure capable of drawing interior timber to the Kitimat facility during the 1960s, Colcel wrote a similar story of expansion and integration to the north. New sources of Scandinavian capital played a part here too, involving a partnership with Svenska Cellulosa Aktiebolaget of Sweden to build a 750-ton bleached kraft mill next to the existing sulphite plant at Prince Rupert. After a dismal first decade that had seen Colcel pile up losses of $14 million, the company had opened the interior's first pulp mill at Castlegar in 1960, overcoming problems in integrating the two operations to record "modest profits" during the early 1960s.[89]

Williston's timber award provided the basis for further expansion in the northwest, where Colcel had cut a wide swath through TFL No. 1 under a liberal interpretation of sustained yield principles. Commencing with an allowable annual cut of 14.5 million cubic feet based on initial "crude estimates" of timber volumes and quality, Colcel had negotiated a steady increase in harvests conditional upon expansion of plant capacity. Between 1952 and 1964 the sulphite mill's output rose from 200 to 530 tons per day, producing a corresponding increase in the annual allowable cut (AAC) to 35 million cubic feet over the same period. Although more precise inventories and improved forest management provided the technical legitimacy for rising harvests, the more practical rationale rested upon a government-industry accord that adjusted logging rates to the capacity of booming markets. By 1960 the total area clearcut by coastal operators reached some 60,000 hectares a year, yet the Forest Service and TFL holders planted less than 8,000 hectares annually. Between 1955 and 1972, allowable annual cuts on coastal TFLs rose by 91 per cent per productive acre. Province-wide, actual cuts on TFLs went up by an astonishing 301 per cent per productive acre during the same period.[90]

Colcel was not alone in playing the sustained-yield game in a way that elevated private over public interest, then, abetted by a succession of forest ministers who confused headlong growth with sustainable development. The bill for such policies would not come due for some time, however, and the announcement of another mill for Prince Rupert seemed to herald another advance in realizing the northwest's industrial potential. Svenska Cellulosa took a 40 per cent interest in the new Skeena Cellulose, planning to market the pulp in Europe, Africa, the Middle East, Mexico and South America. Central to the new enterprise was Williston's timber award, which took the form of a new TFL rather than expansion of the existing licence. Skeena Kraft received TFL No. 40 in February 1965, its 22 million cubic foot AAC sufficient to supply about 40 per cent of the new mill's requirements. A subsequent amendment of its boundaries saw the firm give up timber in the Kalum valley about 50 miles north of Terrace for a strip of Crown forest in the Nass River valley, making TFLs No. 1 and No. 40 a continuous area of about 8,550 square miles.[91]

Colcel financed new housing subdivisions at Prince Rupert for the estimated 350 workers required by the new mill, which entered production at the end of 1966. A new logging subsidiary, Twinriver Timber, supplied both mills from the Terrace headquarters. Nass Camp, equipped with trailers for family housing, served as the focal point for logging in the Nass valley, where the Nisga'a made up 25 per cent of the labour force. By 1967 Twinriver had pushed its main logging road 70 miles north from Terrace, nearing Meziadin Lake, while on the north side of the Nass, construction of the provincial government's Stewart-Cassiar road continued. Twinriver operated four log dumps on the upper stretches of the Nass, the company's 20-truck fleet and another 25 contractors hauling up to 20 miles to those points. Fluctuating water levels and spring salmon runs sometimes interrupted log-driving on the shallow, fast-moving Nass, requiring the construction of a hundred "fin booms" to channel logs away from beaches and sand bars. By the early 1970s crews in aluminium jet boats inspected the booms, broke up small jams, and towed front-end loaders and cats on barges to sandbars when larger obstructions developed.[92]

Colcel also expanded its Skeena River operations during the 1960s, buying Prince Rupert Sawmills in 1965 and acquiring about 20 timber sales in 1967 from Hazelton Sawmills despite poor financial performance caused by high Skeena Cellulose start-up costs and a lengthy strike at Castlegar. After 1967 losses of $4.1 million, a major shift in corporate structure the following year saw Celanese sell part of its 87 per cent interest in Colcel to Svenska Cellulosa. At the same time, the Swedish firm exchanged its equity in Skeena for a 41 per cent share of Colcel. But losses in 1968 hit

$9.6 million, and early in 1969 the Celanese Corporation announced its intention to dispose of its remaining 49.9 per cent interest in Colcel. A small profit in March 1969 represented Colcel's first in 28 months, but shareholders would have taken meagre solace in the $1.5 million first-quarter loss. A strengthening pulp market brought a $1.4 million profit by the end of the year, the first since 1965. Fresh off this triumph, Colcel continued to consolidate its fibre supply, buying the Pohle Lumber Company later in 1969. In addition to the Terrace sawmill, Colcel gained Pohle's quota in the Skeena PSYU.[93]

Construction of a sawmill and chipping facility at Kitwanga in 1970 expanded the range of Colcel's integrated structure still further, but over an enormous geographical area. Trucks carried sawlogs from the Nass River valley 50 miles to the Kitwanga mill and 70 miles to the Pohle plant at Terrace. The CNR continued to haul logs and chips from the Skeena River operations to Watson Island. Despite a $1.5 million profit in the first quarter of 1970, Colcel laid off some of the 500 loggers employed by the Twinriver subsidiary.[94]

Mounting pollution problems at Prince Rupert contributed to the firm's difficulties. Pressure from commercial fishermen over fish kills had forced Colcel to construct a $2 million effluent pipeline in 1968 to disperse waste from the Watson Island mills, but a break in the two-and-a-half-mile line in 1970 caused a two-month shutdown of both mills. Fined $3,000 later that year under the Federal Fisheries Act for discharging wastes into the waters near Prince Rupert, and facing attacks by Skeena MP Frank Howard, the firm defended its record in environmental protection. Mill wastes had adversely affected some small bodies of water, an executive conceded, but the company had embarked on a program to convert the old sulphite mill to an ammonia-based process, and would work with governments to meet pollution-control standards.[95]

Growing public concern over Colcel's environmental impact exacerbated the firm's financial distress. Indeed, by the end of the 1960s the entire industry had begun to feel the pressure from a new movement devoted to environmental preservation. Across North America an increasingly affluent, urban, educated and mobile population had come to appreciate nature more for its aesthetic value than its commodity potential, generating demands for policies that would provide for ecological integrity, a better quality of life and greater public participation in the regulatory process. The Social Credit party held on to power in 1969, but that same year gave rise to the Scientific Pollution and Environmental Control Society (SPEC) and a provincial chapter of the Sierra Club. Greenpeace came into existence in 1971. The closed policy-making community comprised of corporate executives and government officials that had pushed the annual harvest from 22 million cubic metres in 1950 to almost 55 million by 1970 dug in to defend the legitimacy of sustained-yield forestry.[96]

7 Winding Down

Up-Coast Forests and Communities
After the Boom

The industry-government accord in postwar forest exploitation gradually unravelled in the years after 1970. Part of the explanation lies in the failure of sustained-yield management to deliver on its own promises. Mill closures, rising unemployment among forest workers, and community dislocation left a legacy of disappointment and cynicism in many areas of the province, and the up-coast region is no exception. Other sources of stress can be found in the rise of new social values that prize forests for what they offer humanity outside the marketplace, and in First Nations' persistent pursuit of just land-claims settlements. Regional discontent with an economic system that enabled outside corporations to consume up-coast resources, with no apparent regard for regional well-being, also wore away at the legitimacy of policies set in Victoria and Vancouver. It is the recent intersection of these various discontents, in the context of a global economy that no longer offers a secure place for British Columbian forest products, that brought into question sustained yield's rigid, top-down management style.

The New Democratic Party (NDP) government of Dave Barrett confronted the implications of a faltering postwar boom and British Columbia's ecological awakening in the early 1970s. When Crown Zellerbach and the Celanese Corporation prepared to sell off the Ocean Falls and Prince Rupert pulp mills, the NDP bought the plants and operated them as Crown entities. New regulations to fulfil multiple-use promises also came into effect, both initiatives earning Barrett industry condemnation. The appointment of Peter Pearse to head a Royal Commission gave corporate executives an opportunity to defend their management practices, but also provided a forum for First Nations and community groups to express disenchantment with TFL administration.

By the time Pearse submitted his report, Bill Bennett and the Social Credit party were back in power, and while their 1978 Forest Act only entrenched the power of dominant firms over timber supplies, it also acknowledged demands for a more balanced approach to forest planning. As the Ocean Falls and Prince Rupert operations floundered and a deep

recession destabilized the wood-products economy, the postwar boom officially came to an end. The government closed Ocean Falls permanently in
1980 and managed to sell off the Prince Rupert holdings a few years later.
Several of the major companies followed suit, leaving for more attractive
investment opportunities elsewhere. The timber-supply picture contributed
to the erosion of optimism. Estimates revealed that the resource base would
permit no future expansion in many coastal areas.

The once unchallenged Forest Service faced increasing pressure from
First Nations and environmentalists on the Queen Charlotte Islands. Cam-
paigns for the preservation of coastal watersheds erupted, as activists in
Hazelton and Bella Coola pressed for a new community-based alternative
to corporate forest management. Perhaps nothing represented the failure of
the old order more clearly than events at Prince Rupert and Terrace, each
so dependent upon whichever interests controlled the province's first TFL.
Nisga'a protests over operational standards on the licence gained legitimacy during this period, backed up by studies that exposed shoddy practices
whether under public or private ownership.

The massive Prince Rupert pulp enterprise now gathers rust after the
most recent NDP government's bail-out. And history has repeated itself as
the Liberal government headed by Gordon Campbell places faith in a saviour from the private sector. Part of TFL No. 1 now belongs to the Nisga'a,
however, and elsewhere along the central coast First Nations have drawn
together in asserting their claim to what has become known as the Great
Bear Rainforest. The Campbell government has begun acknowledging
aboriginal resource rights in the form of timber and revenue-sharing agreements. It is far too soon to draw conclusions about these accords, much less
to assess the long-term social and economic implications of regional land-
use planning initiatives for the central coast. From a historical perspective,
it is perhaps sufficient to assert their departure from long-standing patterns
of resource allocation and planning.

W.A.C. Bennett's Social Credit government held fast to its promotional
ethic after British Columbians returned it to office in 1969. Relations
between Minister of Lands Ray Williston and the Council of Forest Industries, the trade association that enabled large firms to speak with a single
voice, had settled into a comfortable policy-making network. "Within the
free enterprise economy, I do not think any country controls its basic raw
resources in any more positive a fashion than we do in British Columbia,"
Williston declared in 1969. Notwithstanding frequent bickering over Forest
Service stumpage appraisals, industry and government shared a basic
consensus on the need for secure, long-term tenure to attract and retain
investment capital.[1]

But tension over forest policy grew in the early 1970s as the dispute
over the Nitinat Triangle on the west coast of Vancouver Island hinted at

the land-use conflicts that would come to envelop the province. The NDP was also making waves, criticizing the TFL as a give-away of public resources to multinational corporations. Opposition forestry critic Bob Williams called the tenures "Banana Republics". The Social Credit government established a cabinet-level Land Use Committee as a planning instrument under Williston, who championed the multiple-use concept of forest management and called on industry leaders to join with government in a campaign to boost public confidence.[2]

During the summer of 1972 the Social Credit party had more immediate campaign issues to contend with: a battle for re-election with a more vigorous New Democratic Party, led by Dave Barrett. Barrett pledged to increase government revenues from the forests without slowing economic development, criticized the TFL as a symbol of Social Credit generosity to corporate supporters, and promised a more balanced approach to land-use planning. After 20 years under W.A.C. Bennett's leadership, British Columbians agreed with Barrett's "Enough is Enough" campaign slogan, giving the NDP a solid majority in September 1972.[3]

After the euphoria faded, Premier Barrett and Minister of Lands, Forests and Water Resources Bob Williams, faced a sobering reality in the central and north-coast forestry sector. Crown Zellerbach had begun de-investing at its older Ocean Falls plant in favour of the new Elk Falls facility during the 1960s. In 1963 the firm shut down its original kraft machine, barging surplus production to a new unit at Elk Falls. The downsizing continued in 1967 with the closing of "outdated, small and uneconomical" sulphite and kraft mills, forcing the layoff of 120 workers. Crown Zellerbach's "consolidation programme" at Ocean Falls gained momentum in 1968, when another 150 workers received pink slips. The spectre of costs associated with new air and water pollution-control regulations, along with the expense of cleaning up accumulated waste, no doubt contributed to Crown Zellerbach's desire to abandon Ocean Falls. Power shortages caused by a succession of dry years provided further incentive to shift production to the modern $100 million Elk Falls plant.[4]

Crown Zellerbach president Robert Rogers informed the "rain people" of Ocean Falls of their fate in April 1972, announcing that the facility would close permanently the following March. Rogers cited obsolescence and geographical isolation as the primary reasons for the closure, and Williston eased the way in one of his final ministerial decisions, permitting the firm to transfer timber from its central-coast pulp leases to Elk Falls. As Ocean Falls emptied after turning out its final roll of newsprint on August 31, 1972, the election of the NDP created a glimmer of hope. Nationalization of the industry had been an old policy of the Cooperative Commonwealth Federation (CCF, the predecessor of the NDP); perhaps the new government would save the town.[5]

While Williams and Barrett pondered the Ocean Falls situation, prospects at Prince Rupert took an equally bleak turn. After announcing a

$4.5 million pollution-control plan in 1971, and completing a $9 million expansion of its northwest mills early the following year, the Celanese Corporation abruptly announced its intention to sell Colcel. Having logged the best and most accessible trees from its TFLs amalgamated in 1968, and now facing the prospect of higher costs involved in moving logs from the tip of its licence on the upper Nass River to Prince Rupert, Celanese was prepared to walk away from its sustained-yield commitments.[6]

Williams responded to this crisis in the coastal economy in dramatic fashion, purchasing the Prince Rupert and Ocean Falls operations along with two interior plants between March 1973 and February 1974. Driven "more by circumstance and pragmatism than by any grand nationalization agenda," Jeremy Wilson argues, Williams had moved to reclaim Crown forestland from outside interests, save industries vital to coastal communities and make government an active player in the industry. The NDP created Canadian Cellulose (Cancel) to take over 79 per cent of Colcel's northwest and Castlegar assets, assuming $70.5 million in bond and bank debt while Celanese wrote off its subsidiary's $73 million debt to the parent company. Williams had earlier blocked Celanese's attempt to sell the profitable southern interior operation to Weyerhaeuser, on the grounds that it would have made finding a buyer for the debt-ridden Prince Rupert facilities difficult.[7]

The major problem in the northwest was the deteriorating sulphite plant, plagued by pollution-control problems, corrosion and poor markets. The newer kraft mill could be expected to make money given reasonable wood-supply costs, but that demanded a greater supply of chips from regional sawmills. Only about 20 per cent of the raw material processed by the two Prince Rupert mills came in the form of residual chips, meaning not only high wood costs but waste of lumber-quality logs in pulping. The efficient Terrace and Kitwanga sawmills acquired by Cancel provided a partial chip supply, but new president Ron Gross needed more chip-purchasing agreements to bring down fibre costs.[8]

But Prince Rupert's fibre needs raised another confusing and uncertain situation. By the early 1970s the Prince George pulp mills were shipping chips by rail and truck from as far west as Houston. The Buckley Valley Pulp and Timber enterprise there had previously fallen under the control of Consolidated-Bathurst of Montreal and Bowater Paper of London, who failed to deliver on promises to build a pulp mill. The owners established a modern sawmill at Houston in 1970, but the operation failed miserably. Northwood Pulp and Timber bought the mill in 1972, and began feeding its Prince George pulp mill with the chips. Prince George Pulp and Paper was also desperate for chips, setting the stage for a struggle between the coast and central interior for access to needed sawmill residues. The question of whether the chips from the northwest would go east or west remained to be answered. For the moment at least the region felt "more secure and confident", said Terrace mayor Lloyd Johnstone.[9]

By early 1973, as Williams negotiated with Crown Zellerbach for a possible government takeover, only about 300 residents remained in the once-bustling town of Ocean Falls. Their persistence paid off in March, when Williams announced the $1 million purchase of the mill and town site. A separate Crown corporation, the Ocean Falls Corporation, would be established to operate two newsprint machines, with Crown Zellerbach providing logs at market prices until 1975. The plant's financial viability was uncertain, Williams admitted, but the government considered it imperative to maintain the central-coast community, and would accept short-term losses while exploring opportunities. [10]

A $3 million loan enabled the Ocean Falls Corporation to commence production, and a reported 1973 loss of $850,000 represented quite an accomplishment given interest payments and start-up costs. But studies commissioned by the NDP exposed the operation's uncertain future. Cheap hydro power and abundant timber had drawn industry to Cousins Inlet, but the former would not support plant expansion, and the routing of logs from the pulp leases to Elk Falls deprived Ocean Falls of the fibre needed to establish a proposed kraft mill. No helping hand could be expected from an industry that loathed the NDP's acquisition program, but a small 1974 profit, thanks to a healthy world pulp-and-paper market, quieted the critics temporarily. Meanwhile, Williams tried to interest private investors in economic diversification at Ocean Falls as part of a larger central-coast development strategy. [11]

Declining opportunities in fishing and cannery work contributed to the central coast's economic problems, especially among First Nations. Fishing became more capital-intensive after World War II, and licence limitation under the Davis Plan of 1968 drove the cost of participation beyond the reach of many aboriginal fishers. Postwar concentration in the canning sector, a process that saw plants in outlying areas close in favour of large, diversified operations at Vancouver and Prince Rupert, deprived hundreds of aboriginal women of seasonal work. B.C. Packers closed its plant at Klemtu on Swindle Island in 1969, displacing 93 villagers, about one-third of the band's population. Bella Bella band members also felt the effects, and when the Namu cannery ceased operation in 1971 even seasonal employment became difficult to find. All of the Rivers Inlet canneries had closed by the early 1970s, leaving the Milbanke plant at Shearwater the only source of fish-processing work for central-coast First Nations. [12]

The central-coast forest industry, on the other hand, provided "considerable room for expansion" during this period. Crown Zellerbach and other operators in the Bella Coola region had confined their logging to the valley bottoms and A-framed the better stands on the outer coast. That pattern left considerable volumes of low-quality cedar and hemlock, but largely dispersed in small drainages and on steep slopes along the inlets. Past clearcutting had eliminated seed sources on the most productive low-elevation sites, where heavy slash accumulations and brush encroachment made planting necessary, but most cutover areas in the Dean and Rivers Inlet PSYUs showed satisfactory restocking. Local employment opportunities were limited, however, due to the lack of manufacturing outside of Ocean Falls. Central-coast First Nations faced special problems in securing forest-industry jobs. Most males had experience only in fishing and would require training in logging skills, and the companies found it easier to hire through their Vancouver head offices than to recruit labour from remote coastal communities.[13]

Bitterness between some bands and the forest companies also discouraged aboriginal employment. During the 1960s the Forest Service built a road linking Owikeno Lake, a major Sockeye Salmon spawning ground, to the head of Rivers Inlet. Log booming and towing disturbed spawning habitat, prompting the Owikeno Band to call for protection of a sacred area at the head of the lake and Genesee Creek. While band members worked as loggers on other operations in the region, they did not participate in logging the Owikeno watershed. Instead, the Owikeno Band informed the province of its desire to create a Forest Management and Development Corporation to secure harvesting rights to the watershed. The Forest Service undertook engineering studies of a road from the lake to South Bentinck Arm, a strategy to overcome federal Department of Fisheries objections to the presence of log dumps and towing operations on Owikeno, while promoting access to timber in the watershed.[14]

The Bella Bella Band proposed to secure control over traditional areas through the award of a community TFL when the NDP government commissioned a study of the Bella Coola region. Bella Coola itself requested major sawmill development, a plan Crown Zellerbach had considered and rejected in the early 1970s. The puzzle of up-coast economic development remained unsolved, then. Crown Zellerbach held that the efficiency of log transportation along the coastal waterways, a system that integrated the entire coast with the Vancouver log market, dictated that "most of the unutilized cut from the North Coast region must be converted in the South Coast region."[15]

The authors of the Bella Coola study did not share Crown Zellerbach's analysis, holding out the potential benefits of a new kraft mill at Ocean Falls. "Past forest activity has not helped local people to the extent that it should have," they concluded, but any major development would require careful planning to ensure that benefits flowed both to the region and the province. Skilled labour and technical staff would have to be imported, and problems in recruitment and retention could only be overcome by providing social amenities and better transportation systems. Communities such as Bella Coola and Bella Bella would benefit from increased logging and sawmilling generated by the opportunity to market chips to a new pulp mill.[16]

The desire for forest tenures on the part of the Bella Bella and Owikeno bands was more problematic, but worth pursuing. Both should be encouraged to take part in resource development, if accompanied by education into the capital and managerial requirements needed to avoid bankruptcy. Locally-based companies or cooperatives should receive preference in the award of harvesting rights at Bella Coola and the First Nations settlements, the government study concluded. Failing that, existing operators should give local workers priority in hiring.[17]

While the Ocean Falls problem defied easy solution, and the entire central coast region awaited a policy capable of balancing economic, social and environmental factors, Cancel enjoyed a brief spell of prosperity at Prince Rupert. The firm reported earnings of $29 million in 1974, a source of pride to a government that found itself under a withering attack from industry for its environmental policies. Industry's chief complaint involved the government's Coast Logging Guidelines, introduced in late 1972 to fulfil commitments to integrated resource use. The regulations imposed patch-logging requirements that limited clearcut size, stipulated the retention of deferred areas until cutbacks had restocked, demanded the reservation of leave-strips along fish-bearing streams and imposed tougher road construction standards. Large and small operators alike howled in protest at the cost and alleged inflexibility of the new regulatory regime, apparently gaining a measure of relief in the area of fisheries protection.[18]

Williams' plans to increase the prices industry paid for Crown timber provided yet another source of discontent. But if the entire NDP agenda provoked resentment in a buoyant economy, it would bring outright bitterness when the long postwar boom showed the first signs of winding down. The OPEC oil crisis of 1973, inflation, and slumping lumber and pulp markets by the end of 1974 put Williams on the defensive. He reduced stumpage rates, and in the summer of 1975 appointed Peter Pearse to head a Royal Commission to investigate the entire structure of tenure and forest policy.[19]

Barrett and Williams had other problems to contend with that summer, ones set in the colonialism of the province's history and now showing signs of erupting. The Nisga'a blocked Cancel roads through three Nass River valley reserves for a week that spring in an attempt to force the provincial government to negotiate their land claim. That event drew little public notice, but on the Queen Charlotte Islands resistance to industrial forestry was building among the Haida in Skidegate, Queen Charlotte City and Masset. The Skidegate and Masset bands had organized the Council of the Haida Nation during the early 1970s to advance their claim to the islands. What became known as the South Moresby conflict emerged into the public eye in 1974 in association with Rayonier's TFL No. 24, when the firm began preparations to move contractor Frank Beban's operation from Tekuakwan Island to Burnaby Island. The Skidegate Band, seeking to protect a valued site of food-gathering, raised the issue with Rayonier and Forest Service officials.[20]

Barrett visited the Charlottes that October, agreeing to place the area under a five-month moratorium. When Rayonier announced that it would move Beban to Lyell Island instead, the controversy appeared settled to the Haida's satisfaction. But Rayonier still planned to log Burnaby Island eventually, and remained uncertain about the government's ultimate intentions. Then, late in 1974, an organization comprised of young, recently arrived Americans called the Islands Protection Committee joined forces with a group of Haida to call for the preservation of a wilderness area on Moresby Island, a proposal that encompassed 40 per cent of the area designated as Rayonier's TFL. At this point the idea lacked official Haida support, but as Evelyn Pinkerton relates, forest management and land claims issues became increasingly linked over the 1974–76 period.[21]

The opening act in the Moresby drama unfolded as Peter Pearse began Royal Commission hearings around the province. The process provided TFL holders with an opportunity to defend their management record and press for the tenure security they considered essential to investment and long-term planning. Rayonier's brief described the TFL as the best tenure yet devised for Crown land management, providing security for investment and enlisting the "responsibility, energy and capital of the licensee both in the manufacturing process and in improved forest management to the mutual benefit of the licensee and the public." Independents called for eviction of the majors from the PSYUs, but few within industry called for dramatic changes to the TFL tenure. Even the Truck Loggers Association praised forestry standards on the TFLs, and cited the opportunities afforded independents under the contractor clause in supporting continuation of TFL management.[22]

While industry players closed ranks against environmentalists, deplored the ruinous cost of NDP regulations and played up the obligations assumed under TFL contracts, outsiders advanced a quite different set of views during Pearse's visit to Prince Rupert. Perhaps the most radical perspective originated on the Queen Charlottes, in a Skidegate Band Council brief which criticized Sloan's assumption that cooperation between government and corporations in forest management would embrace both private and public interests equally. "To suggest, as Judge Sloan did, that the people of the Province should safeguard the timber resource for centuries in the warm heart of the large companies is, we submit, being naïve and blind to the actual operations of the multinationals," the Skidegate Band asserted.[23]

The Haida had utilized the forest for thousands of years before the term "sustained yield" had been coined and the corporations arrived, the brief continued. Many now earned wages working on MacMillan Bloedel's TFL No. 39, but were "denied the benefits realized from the profits of their labour". Moreover, that licence, encompassing seven blocks on Vancouver Island, the mainland and the Charlottes, defied Sloan's emphasis on regional sustained yield. What could prevent the company from cutting out its Queen Charlotte holdings completely in the future, so long as the cut from the entire tenure did not exceed ministry limits? The best safeguard for sustained employment and "respect of the forests" was community-based resource development, meaning village ownership of TFLs. Only then would the community provide work and services needed to realize "the benefits and pleasures of life presently only realized by the affluent southern citizenry".[24]

Percy Gladstone and Carey Linde appeared before Pearse, who devoted considerable attention to the village-controlled TFL concept. The central idea, they explained, was for the Skidegate band to apply to take over part of the three Queen Charlotte licences when they came up for renewal. The Masset Haida and non-aboriginal communities on the islands could also benefit from such a policy. Pearse, raising the issue of resource-poor communities, asked if the existing principle of Crown ownership and revenue redistribution would not be more equitable. "That is all very well in theory," Gladstone responded, but not "if you lived in the Queen Charlottes and saw the barge-load after barge-load of logs going out with very little coming back into the village, into the communities." Sixty miles of blacktop and a ferry represented the only visible returns after over 50 years of logging under the present system, he observed.[25]

A commission official went on to suggest that meeting the band's proposal would require the government to take timber away from TFL holders, a step that might conflict with the Crown's legal obligations to licensees. The application would be filed when the tenures came up for renewal, Gladstone responded. A MacMillan Bloedel representative carried this line of inquiry further, asking if the band council intended to compensate the licence holders for lost timber and improvements. Gladstone replied that the Haida land claim took precedence, concluding: "The Haida people feel that they should be compensated before anyone asks us for compensation."[26]

The Skidegate Haida were not the only northwest group to express an alternative vision to the Pearse Royal Commission. The Nass Valley Communities Association, a loosely organized group of non-Nisga'a residents, expressed their views on life within TFL No. 1 and changes needed to make the tenure serve the region's people. "We are the people who live every day in a Tree Farm Licence," representative Wayne Samland told Pearse. "We work in it, we drink the water from it, we breathe the air, we hear it, we sleep with it, it is all around us, we drive the roads of it."[27]

The association's basic complaint concerned the devotion to corporate profit that elevated wood production above all other land and water uses in the valley. The Nass fishery received some official attention because of its commercial value, but wildlife, agriculture, community development, water quality and "the future" were protected only as an afterthought by understaffed, poorly coordinated public agencies from distant offices. Even the Nass River salmon runs fared poorly under this regime due to the log drives that disturbed spawning grounds and clearcuts that typically extended to riverbanks. The federal fisheries officer felt "lucky to get concessions on even minor logging practices", reflecting the shortcomings of a planning process controlled by an outside corporation for the benefit of people who lived elsewhere. "Regional benefits are marginal and control non-existent," the association declared in its brief.[28]

The proposed solutions included settlement of the Nisga'a land claim, integrated resource management through the creation of a local planning team to "balance the power and control of the operating company", and the return of stumpage revenues to the area to promote diversified, environmentally friendly economic development. Expressing "the concrete dilemma of daily life in our valley", the Association voiced the resentment of an area treated by corporations and governments alike as "a remote, unpeopled fund of exclusive timber dollar potential".[29]

More practical issues relating to the short-term orientation of Cancel and other large companies also emerged out of the Royal Commission proceedings. The Kispiox Valley Community Association joined the Nass Valley people in calling for a reduction in the size of clearcuts, less waste and a recalculation of Cancel's allowable annual cut (AAC). The new firm and its predecessor had concentrated logging in the southwestern portion of the TFL, taking advantage of an allowable cut based on the volume of the entire area to deplete the more accessible part of the claim. That left it in control of vast forests within both the TFL and the region's PSYUs that might be better utilized by small sawmills, "the only diverse elements" in the Kispiox valley economy. Moreover, the cutting pattern would eventually force local residents to move in order to keep their jobs as operations shifted north, a trend already apparent.[30]

The Kispiox brief included references to inadequate reforestation and the loss of job opportunities to mechanization, described clearcutting as "at best a necessary evil" and called for wider leave-strips along streams. And it shared with the Skidegate and Nass submissions a community's deeply-rooted concern for the full range of benefits the forest could provide for present and future generations. "The forest companies may move on and the government may ignore us," said the Kispiox document, "but we will still be here, and our children will be here, and hopefully their children will be here.... We want to use and enjoy our forest and we want our children to be able to do the same." At the heart of these presentations lay disenchantment with government policies that bred concentrations of corporate power, and the regional economic inflexibility that such a structure produced. Making wood available to rural settlers in addition to the reserves required by large companies would restore options and create a more resilient economy. Finally, community involvement in Kispiox valley forest management would consider the local knowledge and social values of people who must live with the consequences of "environmental damage, depleted forests, or decaying one-company economies".[31]

While Pearse went on to Vancouver to hear the presentations of major tenure holders, Dave Barrett issued an election call. By this time Bill Bennett had succeeded in drawing prominent Liberals into the Social Credit party to unite the right against the socialists. Barrett's campaign against Pierre Elliot Trudeau's anti-inflation policy earned his party 40 per cent of the popular vote, but the sharp drop in forestry revenues contributed to a dismal economic outlook that played its part in a Social Credit victory. Pearse would submit his report to new Forests Minister Tom Waterland, while Williams sat in opposition. The forest industry, which had regarded his tenure as "more or less ... a reign of terror", rejoiced. Waterland was a "staunch free enterpriser," one industry journal enthused, "with a dislike for any government interference in the marketplace."[32]

Waterland took immediate steps to assist the recession-troubled industry, easing off on enforcement of utilization standards and the Coast Logging Guidelines. That came as welcome news in the Terrace area where the slump, closure of Cancel's sulphite mill for a conversion to kraft production, and winding down of logging in the Kitimat River valley had slashed employment and idled many logging contractors. Addressing the May 1976 convention of the North West Loggers Association, Waterland promised to look into Cancel's dealings with its contractors, but cautioned that some might have to go elsewhere for a few years until the northwest got "back on its feet".[33]

Waterland delayed major initiatives until receipt of the 1976 Pearse report, a wide-ranging discussion of provincial forest policy. The central policy-making trend, Pearse observed, involved the shift of management responsibilities from the over-burdened Forest Service to industry. That had coincided with tremendous expansion, bringing the annual harvest up to 400,000 acres. Thus, a way had to be found to reconcile forestry's traditional production mandate with the public's growing demand for environmental protection. But aside from calling for a more rigorous incorporation of multiple-use principles in forest management, Pearce defended the fundamental principles of sustained-yield forestry. There was no need for a drastic reduction of annual cuts, except in some coastal areas, as lower harvests would only delay the replacement of static old-growth timber by vigorous new plantations.[34]

Pearse, alarmed at the control exerted over cutting rights by a few large corporations, did call for more timber to be made available to small operators. Ten of the province's major companies held rights to 59 per cent of the public forests, pointing to the need for new short-term licences open to competitive bidding by independents. On the other hand, he considered TFLs a great success. The tenure had brought huge areas under sustained-yield management, made it possible for large enterprises to attract investment capital, and holders practised a high standard of management under Forest Service supervision. But one aspect of TFL administration drew critical comment. Harvest rates on the tenures had risen dramatically over the years, Pearse noted, and only a fraction of the increase could be attributed to the growth stimulus companies claimed their advanced forestry practices produced.[35]

Pearse went on to call for adjustments to increase the government's ability to withdraw Crown lands from the tenures for harvest by independents and uses other than timber production, but his overall enthusiasm for the TFL pleased the majors. Waterland appointed an advisory committee to prepare new legislation that would provide industry with the incentive for intensive forest management. The resulting 1978 Forest Act ignored Pearse's call for "modest increases in competitive bidding", Jeremy Wilson asserts, and "removed any residual worries about tenure security left over from the NDP years". The act's tenure provisions, which served to "entrench the concentration of harvesting rights", Schwindt agrees, drew applause from the Council of Forest Industries and denunciation from independents.[36]

Multiple use became official policy under what Waterland called a "new concept of partnership in management between government and industry." The PSYUs were consolidated into Timber Supply Areas (TSAs), requiring the Chief Forester to consider recreational, fisheries and wildlife values in setting five-year harvest levels. TFLs would come up for renewal every 10 years under 25-year "evergreen" terms. Independents would qualify for Crown timber under new Timber Sale Licences, a Small Business Enterprise Program, and the contractor clause of TFL contracts. A range of incentives offered industry financial compensation and increased allowable cuts in exchange for reforestation and intensive silviculture investments. Compensation would also go to major tenure holders if lands withdrawn for parks or other purposes reduced the AAC by more than five per cent. The principle was "good tenure in return for performance and good management", Waterland explained.[37]

TFL holders grumbled about the five-per-cent withdrawal clause, but on the whole expressed approval of the tenure security afforded by the legislation. The return of a healthier wood-products market by 1977 afforded another source of comfort. The Social Credit government, uneasy owners of Cancel and the other properties acquired by the NDP, bundled most up in the new B.C. Resources Investment Corporation (BRIC) as B.C. Timber. Cancel itself had made something of a recovery by this time, cutting its long-term debt in half after the slump of the mid 1970s. The negotiation of long-term agreements with European and Asian customers, while undertaking the sulphite-to-kraft conversion of the original Prince Rupert mill, seemed to revitalize the entire operation. The transition to kraft production would avoid an $80 million pollution-abatement program, provide more efficient fibre utilization, lower per-ton labour costs, and open up new market opportunities, Cancel management claimed.[38]

The conversion, completed in October 1978 at a cost of $135 million, restored employment in the entire Prince Rupert complex to 760 workers while bringing daily bleached pulp production to 1,100 tons. But Cancel's fibre supply problems would not go away. By this time the Twinriver subsidiary handled 45 per cent of the logging, contractors supplying the balance. Those with Cancel seniority lived in Terrace, while loggers with less service were assigned to Nass Camp, a settlement with bunkhouses, trailers, houses for supervisory staff, a bowling alley and a skating rink.[39]

Logging equipment included steel spars, modern grapple yarders and tractors, but ground conditions limited the usefulness of tree shears. Twinriver double-shifted its grapple yarders, their two-man crew complement contributing to reduced labour demands. After peaking in the early 1970s, the population of Nass Camp stood at 120 in 1978. About half of Twinriver's Nass loggers were Nisga'a from the villages of Gitwinksihlkw and Aiyansh. Economics and environmental concerns had curtailed the Nass River log drive, but low-quality timber and long truck hauls continued to plague Cancel's subsidiary. "The decadent hemlock and balsam are the curse of this area," a manager explained. "Because of it we're geared to a pulp economy. It's a real struggle to get sawlogs.... All this, plus high transportation costs, makes for an expensive pulp log."[40]

Cancel had the dubious distinction of being the only major forest products company in the province not to show a record profit in 1978, as the industry enjoyed a brief spell of real prosperity. Earnings picked up in the first quarter of 1979, while company and Ministry of Forests officials discussed reducing the size of Cancel's amalgamated TFL No. 1. The two sides ultimately agreed to delete about 1.7 million hectares of less valuable, remote timber in the Kiteen, upper Nass and Skeena watersheds from the licence, reducing its extent to about 1 million hectares and easing the firm's management responsibilities. A regional timber crunch loomed, and the deletion served the additional purpose of freeing up resources for operators in the North Coast, Kispiox, Kalum and Prince George TSAs. "Most of the independents in the north are a bit short of wood," Ken Berhnsohn noted in 1980, despite government claims that supplies were adequate to maintain production levels. Facing a shortage of accessible timber, Skeena-Price Forest Products sold its Terrace sawmill and cutting rights to Cancel in 1980. The Price Company had acquired Skeena Forest Products in 1969, marketing its chips and pulp logs to Cancel; when it announced plans to sell, Cancel had been forced to step in to protect its long-term fibre supply. That same logic drove Cancel to buy Rim Forest Products of Hazelton, along with its Forest Licence and three Timber Sale Licences, later in the year.[41]

Raw-material supplies had also become increasingly problematic at Ocean Falls, where Ray Williston was back on the scene as chairman of the Ocean Falls Corporation. After recording a $76 million profit in 1976, the operation had faltered badly due to high log costs, interest rates and power shortages. The Forest Service's failure to approve bids on timber in the Rivers Inlet and Dean River areas proved a crippling blow. Losses hit $8 million in 1979, while Williston tried to interest private investment in the operation. For a time, negotiations with the Kruger Pulp and Paper Company of Montreal looked promising, but the Bennett government balked at constructing a $55-million transmission line to carry power from the Alcan powerhouse at Kemano. When the government decided that mid-coast timber supplies would not support Kruger's needs, Ocean Falls was finished.[42]

Williston announced the town's fate on March 6, 1980, prompting an exodus of the 1,100 residents. Relocation expenses and job placement assistance eased the transition for skilled workers. Williston tried to maintain a semblance of community at Ocean Falls, hoping to interest companies in processing the region's low-grade cedar into "flitches" and wood chips there. That would have seen logs cut on two sides and stacked on barges for shipment to Lower Mainland mills for final manufacturing, with the residue chipped for sale on the open market. In the short-term, Ocean Falls might become a focal point of central coast logging, a home base for gyppo operators and loggers employed by the major companies. A long-range vision would have seen the Ocean Falls Corporation participate in establishing a kraft mill to process the mid coast's decadent cedar. A third proposal envisioned an oriented strandboard plant utilizing cedar chips at Ocean Falls. But all of these solutions required private investment, and the deep global recession and high interest rates of the early 1980s made Williston's task impossible.[43]

Meanwhile, a Social Credit government bent on implementing a draconian "restraint" agenda took a dim view of the $500,000 annual expenditure involved in town and plant maintenance at Ocean Falls. Accordingly, in 1983 the Bennett government abolished the Ocean Falls Corporation and made plans to "normalize" the town by demolishing 200 apartments and 160 houses. The remaining 50 residents threw up a blockade when the backhoes moved into action in 1985, but backed off when threatened with criminal charges. The mill equipment went on the auction block, and although the new Ocean Falls Improvement District saved part of the town from destruction, the losses prevented the community from becoming a popular tourist destination or stopping place for Alaska cruise ships.[44]

The demise of Ocean Falls reflected long-term timber-supply problems, a legacy of decades of high-grading and the decision to allow Crown Zellerbach to shift timber from its mid-coast holdings to Elk Falls. The plant had struggled along for a time in the 1970s buying logs on the open market, but escalating chip prices that doubled the cost of pulp logs doomed the operation. Social Credit's predisposition to escape the obligations of public ownership also contributed to the operation's failure, as did the government's reluctance to make the concessions needed to attract private investment. Finally, the early 1980s crisis in the world capitalist economy destabilized the entire forest industry, with repercussions for communities up and down the coast.[45]

Companies had begun feeling the pinch of constricted markets and falling prices by the end of 1979, the beginning of another in the cycle of booms and busts that plagued the industry since its inception. Pulp markets held up better than lumber at the outset, but by mid 1980 all sectors experienced reduced demand The Social Credit government responded with a policy of "sympathetic administration", easing up on utilization standards and reducing stumpage rates. That unleashed a government-sanctioned campaign of high-grading in the Crown forests, and restraint-driven layoffs in the Forest Service further undermined the agency's capacity to monitor industry practices.[46]

As the province's forest administration lost efficiency, disturbing reports of forthcoming timber shortages along the coast circulated. A 1980 report on the Prince Rupert Forest District's Mid-Coast TSA revealed that continued harvesting of the most accessible and productive sites would produce "an abrupt falldown in yield" within 30 years. Exhaustion of the valley-bottom timber would soon force companies into poorer stands on steep slopes and more difficult terrain deeper in the watersheds, where environmental constraints and prohibitive road-building costs dictated skyline or helicopter logging. Insufficient nursery capacity and brush encroachment had hindered the reforestation effort, which in any case offered no solution to the impending falldown. Only an acceleration of spacing and fertilization programs could increase the area's long-term capacity.[47]

Similar studies in other coastal regions confirmed that Pearce's timber-supply predictions had been overly optimistic. In 1983 Waterland announced that the AAC on the Queen Charlotte TSA would be cut by more than half, from 1,074,000 to 450,000 cubic metres. Within a year the Ministry of Forests admitted that many areas would confront falldown under current management practices. The initiation of a five-year, $500 million federal-provincial Forest Resource Development Agreement in 1985 provided an infusion of funds for reforestation and intensive forest management, designed to reduce the estimated 644,000 hectares of not-satisfactorily restocked lands. Critics argued that FRDA and 1987 legislation that increased industry responsibility for forest renewal merely rationalized a provincial harvest rate that continued to rise. The 1990 AAC for Crown lands stood at 74.5 million cubic metres, almost 10 million above the 1985 ceiling.[48]

The Ministry of Forests intended that previously disregarded northern forests would provide fresh sources of supply to maintain the cut. Chief Forester Bill Young set an AAC of 140,000 cubic metres for the Cassiar TSA in 1984, four times higher than the average over the previous five years. The recent closure of the Granduc copper mine had left 500 unemployed, threatening Stewart's existence. Three companies – Orenda Forest Products, Buffalo Head Forest Products and Tay-M Logging – had Forest Licence applications approved, all promising to build processing plants in Stewart.[49]

But none of the companies had initiated construction by 1990, and Tay-M had recently failed to pay its contractors or fulfil financial commitments to the Municipality of Stewart. In place of the prosperity new mills would have brought, Stewart watched the highest quality logs being trucked to Terrace or the port at Kitimat for export to Asia. Caught in a catch-22, the Ministry of Forests permitted the practice while continuing to press for fulfilment of the investment promises. Even the proposed sale of Tay-M's cutting rights to Eurocan promised no relief, as West Fraser intended to move those logs to its Terrace sawmill.[50]

While Stewart officials saw the town's future slipping away, a variety of forces generated a capital flight that saw several of the province's postwar industry leaders depart the scene despite the tenure security they had achieved. Depleted holdings, aging mills, pollution-abatement costs, pressure on the land base from environmentalists and First Nations, the softwood lumber dispute with the U.S.A., and the appeal of fast-growing forests in the southern hemisphere all contributed to de-investment by firms whose mill-sites typically featured "Here Today, Here Tomorrow" signs. Rayonier, which had been absorbed by International Telephone and Telegraph in 1968, left for greener pastures in 1980. Western Forest Products, a new partnership organized by Doman Industries, Whonnock Industries and B.C. Forest Products, picked up Rayonier's assets and quickly closed several mills judged obsolete.[51]

Tom Waterland put Rayonier's departure in the best possible light, hailing the return of local control over the firm's sawmills, Port Alice and Woodfibre pulp mills, and more than 240,000 hectares of TFL tenure. Crown Zellerbach began its withdrawal in 1980, selling its Elk River Timber subsidiary to B.C. Forest Products. That set the stage for a wholesale pullout in 1983, when Fletcher Challenge of New Zealand spent $300 million to acquire all of Crown Zellerbach's Canadian assets. Fletcher Challenge went on to take control over B.C. Forest Products in 1987, enveloping those operations within a multinational structure that included holdings in Latin America, the United States, Australia, New Zealand and South America.[52]

While residents of resource communities around the province pondered the implications of the corporate shake-up, those who depended on pulp manufacturing at Prince Rupert for a living also had cause to wonder about their future. Despite the new kraft mill, Cancel – renamed B.C. Timber in 1981 – encountered high pollution-control expenses just as the bottom fell out of the pulp market. A 1981 $43-million agreement with the province to bring air emission and effluent control equipment at the Skeena Pulp Division up to standards burdened the already struggling enterprise. Only a fair performance by B.C. Timber's Castlegar operation kept the firm's losses to $40 million in the first nine months of 1982. B.C. Timber shut Skeena Pulp down indefinitely on November 26, 1982, adding 1,750 workers to the region's unemployment line. "We have been caught in

Cancel pulp mill at Port Edward, 1981.
BC Archives NA-37137

a cost/price squeeze that is largely beyond our ability to control," President John Montgomery explained.[53]

World pulp prices had dipped to 1975 levels while manufacturing costs had more than doubled since that time. A strong Canadian dollar in relation to European currencies gave Scandinavian producers an insurmountable advantage with European customers who traditionally purchased half of Skeena's output. High wood costs, BRIC President Bruce Howe said, made Skeena extremely vulnerable to any market downturn. The company's over-mature hemlock forests were riddled with interior rot, worthless for lumber, and expensive to log. Citing fibre costs that ran 25 per cent higher at Prince Rupert than at Castlegar, Howe asserted that the value of the northern stands was too low to "justify private investment to cover the required costs of forestry management and road maintenance".[54]

The Skeena Division resumed operation at 60-per-cent capacity in May 1983, and a reduction in stumpage charges helped bring the Prince Rupert plant back to full production in September. But Howe remained pessimistic about the long-run outlook at Prince Rupert, emphasizing the need for lower labour costs and increased productivity to achieve profitability. Meanwhile, the entire Skeena Valley suffered the consequences of B.C. Timber's inability to cope with the slump. Unemployment hit 35 per cent in Terrace during the winter of 1982, when B.C. Timber shut down the

The automatic sorter slashed sawmill labour costs by displacing green-chain workers.
BC Archives NA-38923

town's two sawmills and laid off all the loggers. "People in Terrace are feeling what it's like to be part of a large pulp manufacturing enterprise," a journalist noted, and a survey by the Terrace-Kitimat Labour Council showed that most favoured a more diverse economy capable of sustaining the quality of life offered by the local environment.[55]

But Terrace's environment was showing signs of wear: the valley bottoms had been logged, and the hemlock and balsam stands to the north were suitable largely for pulping only, according to B.C. Timber. The company's plans to rationalize its northern operations threatened to further erode local employment with the conversion of its Terrace sawmills into a chipping plant. The replacement of labour-intensive green chains for stacking lumber by computer-operated sorting systems had already begun to diminish sawmill employment. A $1 million installation at one of B.C. Timber's Terrace mills had replaced 50 green-chain workers, one example of an industry-wide drive to cut labour costs.[56]

The sawmilling sector along the north coast offered even less reason for optimism. Seventy-five-year-old Jimmy Donaldson continued to operate Brown's mill on the Ecstall River, cutting logs supplied by beachcombers and handloggers in a plant still driven by water until a flood took

Brown falls from Brown's mill.
BC Archives NA-10444

out his dam in 1979. Donaldson then converted to diesel at the mill his stepfather founded in the early 20:h century, but planned to repair his flume and revert back to waterpower. The adoption of cardboard packaging at the canneries had curtailed commercial operation at the mill for a time, but Donaldson had resumed activity turning out cannery floats and boat lumber. But no major sawmills operated on the north coast by the early 1980s; Prince Rupert's only mill closed down a decade earlier after falling to profit-cutting low-grade logs. Decadent timber and steep, rocky terrain contributed to high logging costs, and environmental concerns had begun to restrict access to valuable stands.[57]

One such site was the Khutzeymateen Inlet north of Prince Rupert, where the importance of salmon and Grizzly Bear habitat had led to demands that the watershed become an ecological reserve. Anxious to gain access to an estimated $130 million worth of timber and demonstrate the environmental sensitivity of helicopter logging, the Ministry of Forests approved harvesting in the Khuzeymateen by Jack Erickson, who had pioneered the technique in California. Erickson established Silver Grizzly Logging at Prince Rupert in the late 1970s, and began logging without the ecological disruption associated with conventional overhead methods and

Sikorsky S-64 Skycrane.
BC Archives NA-37134

road-building on steep slopes. The incentives offered to Silver Grizzly included the right to export the logs to Japan. The sight of raw logs leaving the north coast was nothing new, of course, but Prince Rupert Mayor Peter Lester hoped to see the region's timber allocated to interests willing to undertake local processing. "Timber from here – our timber – shouldn't go to supply mills in the Lower Mainland," Lester said. "It's about time timber was allocated to businesses that are going to locate here."[58]

Silver Grizzly's experimental helicopter operation in the Khutzeymateen proved unrewarding, and the Ministry of Forests now sought to halt export of the north coast's best logs, broaden the Prince Rupert Forest Region's manufacturing base, and expand industry's access to timber. The regional office put a 125,000-cubic-metre-a-year Forest Licence in the North Coast TSA up for competition, conditional upon the construction of a new Prince Rupert sawmill. Wedeene River Contracting, owned by a Terrace businessman, stepped forward in March 1983 and agreed to have a $38 million, 61,000-board-metre mill up and running by February 1986. Wedeene posted a $50,000 performance bond, purchased land in the Prince Rupert Industrial Park, and began logging at Silver Creek near the harbour with strong support from the community.[59]

But Wedeene had trouble rounding up financing for the mill, and in August 1984 the owner announced that he could proceed only if cabinet granted access to the Khutzeymateen Valley and permitted export of the choice logs. The plan drew the ire of environmentalists, still campaigning to have the Khutzeymateen declared an ecological reserve. Social Credit's new Wilderness Advisory Committee had scientists examine the proposal, asked Wedeene to make its own assessment of the risks logging posed to wildlife, and concluded that Wedeene could log the valley provided that it incorporate field studies on Grizzly Bears into the development plan.[60]

Grants from environmental groups such as the World Wildlife Fund supported studies by bear biologist Wayne McCrory and dissident forester Herb Hammond. While McCrory's work emphasized the "significant and special" quality of the valley's habitat and cast doubt on its capacity to withstand commercial logging, Hammond raised questions about the economic viability of harvesting Khutzeymateen timber. These findings, along with pressure from the environmental lobby, led to a joint study by the Environment and Forests ministries on the potential impact of logging on the bears. If the findings indicated an unacceptable degree of risk, the ministries agreed, logging would be reconsidered. Preservation of the Khutzeymateen would have consequences for north-coast allowable cuts and forest-sector renewals, but environmentalists argued it would produce no mill closures or lost logging jobs. Given that the Committee on the Status of Endangered Wildlife in Canada had recently designated the Grizzly Bear a threatened species, Mel-Lynda Anderson concluded, preservation would be a "politically safe move since the human cost … is relatively low".[61]

In 1988, Wedeene completed its Prince Rupert Sawmill, capable of cutting only small logs, but the government slapped a heavy tax on log export profits early the following year. The company, which had been sawing only half of its Forest Licence harvest, sending 38 per cent to the Vancouver market and exporting the rest, went into receivership in September 1990. Wedeene wanted to boost its exports, and got City of Prince Rupert officials to support a second licence capable of almost doubling its annual cut. Forests Minister Claude Richmond denied the application and freed the company of its obligation to operate the mill in order to preserve 57 logging jobs, but the concession did not bring the company out of receivership. Closure of the mill put 69 employees out of work, and the province placed Wedeene's assets on the market on the condition that the buyer process logs from the Forest Licence in Prince Rupert.[62]

West Fraser acquired Wedeene's small plant, logging equipment, and cutting rights in 1991, two years before Mike Harcourt's new NDP government placed the Khutzeymateen valley under a permanent logging moratorium. Forests Minister Dan Miller said that the absence of cutting rights there made the decision an easy one, but West Fraser officials warned that its withdrawal from the North Coast TSA would reduce the annual cut and result in eventual job losses. West Fraser went on to build a new high-tech Prince Rupert sawmill that began production early in 1995. The company received some 700 applications for employment at its $25-million North Coast Timber operation, a measure of the region's staggering economy. The new mill emphasized workforce flexibility in an effort to make a success of log processing at Prince Rupert, rejecting mass production in favour

of cutting both dimension lumber and "high-end products" to the specifications of Pacific Rim customers. The shift to flexible production in sawmills and pulp mills after the recession of the early 1980s reduced labour
requirements dramatically throughout the industry, but Prince Rupert welcomed North Coast's 45 new jobs during a time of great uncertainty over
environmental issues, land claims and the town's major pulp enterprise.[63]

By 1985 the value of BRIC shares had fallen by about half, and B.C.
Timber was going by the name of Westar Timber. The new name meant
no departure from the clearcut logging conducted by predecessor companies, but the Nisga'a had made some progress on the land-claim front over
the previous decade. After the Supreme Court of British Columbia denied
the validity of aboriginal title in the Calder case, a ruling upheld by the
Court of Appeal in 1970, the Nisga'a appealed to the Supreme Court of
Canada. There they achieved what Hamar Foster calls a "remarkable but
complex victory". Six of the seven judges affirmed the existence of aboriginal title based on occupation, although three of them held that such title
had been extinguished by colonial homesteading laws. The seventh judge
ruled that since under British Columbia law a plaintiff had to obtain consent to sue the Crown, and the Nisga'a had not, the case should never have
been brought before the court. The tie meant a technical loss for the
Nisga'a when the court handed down its decision on the Calder case in
February 1973. But six judges had acknowledged that aboriginal title to the
land existed under English law when the colony of British Columbia was
founded. Three of them had gone further to argue that Nisga'a title had
never been extinguished.[64]

The federal Liberal government of Pierre Trudeau quickly initiated a
policy of settling land claims where treaties had not extinguished such title.
The Nisga'a Tribal Council, now led by James Gosnell, began negotiations
with the federal government in January 1976. But the province refused to
take part in the negotiations, maintaining its denial of aboriginal title or any
interest in land based on use and habitation. Not until 1990 would the
provincial government sit down at the negotiating table.[65]

While the Nisga'a negotiated, they also launched an effort to have B.C.
Timber's TFL No. 1 revoked so that they could obtain their own Nass-
valley licence. In 1982 the Nisga'a Tribal Council hired Herb Hammond
of Silva Ecosystems Consulting to scrutinize the forest practice record on
the tenure. When Hammond presented his report in 1985 B.C. Timber had
become Westar, but his conclusions suggested that, from its inception, TFL
management had been oriented to short-term profit maximization rather
than sustainability. Artificial and natural reforestation of the most productive lands nearest the mouth of the Nass River had achieved low levels of
restocking, and heavy brush dominated many sites. Between 9,000 and
15,000 hectares of not-satisfactorily-restocked land existed, much of it on
the lower Nass River, where periodic flooding complicated stand re-
establishment.[66]

Hammond was particularly critical of the high-grading that had occurred after the 1970 amalgamation of TFLs No. 1 and No. 40, a development accompanied by a sudden one-million-cubic-metre increase in the AAC. Despite an apparent Forest Service requirement that the more remote forests from the newer licence contribute to the elevated harvest, Colcel and Cancel had continued concentrating the cut in the lower Nass River's larger and more cheaply accessed bottomland timber between 1970 and 1978. Manipulation of the AAC in conjunction with slack utilization practices, poor reforestation and inadequate attention to soil protection left the Nisga'a a legacy of high unemployment and environmental degradation.[67]

Westar management discounted the Nisga'a claims even before reading the report, announcing negotiations with the Forest Service for a further relaxation of utilization standards in recognition of the region's decadent forests. All operators in the Nass had suffered "substantial losses, subsidizing the local economy, with no return to investors", Westar's Sandy Fulton declared. Another report by provincial Ombudsman Karl Friedmann, initiated by Nisga'a criticism of the Ministry of Forests, confirmed that the government had failed to properly enforce reforestation obligations on TFL No. 1 in order to encourage profitable operation. Backed into a corner, the ministry reduced the area of the licence by about one-third in 1985. Forests Minister Tom Waterland explained that Westar had been unable to make full use of its allowable cut, and threw the deleted lands open to Forest Licence applications.[68]

Yet another study commissioned by the minister, this one conducted by Harry Gairns of Industrial Forestry Service, noted the absence of firm reforestation commitments in Westar's management plans. But by the time it appeared, BRIC had sold most of its Westar Timber Division assets to Repap Enterprises of Montreal for $100 million. Included in the 1986 transaction were the two Prince Rupert pulp mills, the Terrace sawmill, and much of the TFL. Westar retained the Hazelton sawmill, but welcomed the opportunity to withdraw from the pulp business and lower the corporation's debt. Repap received a $75-million loan from the British Columbia Development Corporation, pledging a $40-million investment to modernize the Skeena operation in pursuit of lower costs and productivity gains. For the Nisga'a, then negotiating with Westar and the federal government to purchase TFL No. 1 or have at least the lands recently relinquished by Westar transferred to their control, the deal came as a betrayal. Their plan for the licence featured the establishment of a nursery, a plant to produce finished products, and the creation of 200 new jobs in logging, forestry and manufacturing. Even a partial fulfilment of that vision would have represented a vast improvement over the carnage left behind by Repap. Nor did the Nisga'a succeed in obtaining the lands recently deleted from TFL No. 1, which the ministry awarded to Terrace and Vancouver contractors.[69]

For an increasingly belaboured Ministry of Forests, its reputation tarnished by the revelations surrounding management on TFL No. 1, the

Repap deal at least seemed a triumph for old-fashioned commodity production values. Elsewhere along the coast, and on the Queen Charlotte
Islands, victories were hard to come by during the 1980s. On the Charlottes, conflicts with the federal Department of Fisheries, environmentalists
and the Haida placed increasing strain on industry's freedom to drain timber to southern mills. A stronger federal Fisheries Act passed in 1977 set the
stage for conflict with the province and its forest industry. Adoption of a
$150-million salmon enhancement program to boost a sagging fisheries
sector added urgency to habitat protection concerns, and made for increasing suspicion within forestry circles that loggers bore too much blame for
depleted salmon stocks.[70]

The first shots in the fish-forestry war were fired at Riley Creek, a
salmon stream in the Rennell Sound area of Graham Island. There Q.C.
Timber, a Vancouver-based subsidiary of a Japanese firm, ran afoul of
Fisheries Officer Jim Hart late in 1978. Concerned that logging of the steep
slopes above the creek would trigger slides and destroy spawning grounds,
he ordered the firm not to proceed with operations along a 16-hectare area.
Ironically, the stop order came just after Fisheries Department official Wally
Johnson had described a new era of understanding between the two indus-
tries at the annual Truck Loggers Association convention in Vancouver.[71]

Forests Minister Tom Waterland turned a minor irritant into a major
confrontation in his address to the same meeting, making public a telegram
to Fisheries Minister Romeo Leblanc. Calling the logging ban "a pursuit of
a single minded objective, made without regard to the cost factors involved
to the company," he demanded cancellation of the order. Ottawa did not
heed Waterland's demand, and the Forest Service authorized Q.C. Timber
to log the tract. When two weeks of bickering between the agencies failed
to resolve the dispute, Q.C. Timber decided to defy the federal order. "We
have to bring things to a head," Q.C. Manager John Sexton explained, and
on March 19, 1979, fisheries officers arrested members of a Q.C. logging
crew. The company responded by bringing in more loggers, prompting further arrests. By mid-week eight employees had been arrested, and the confrontation showed no signs of ending. "Our instructions are to keep arrest-
ing," Hart asserted. "Theirs are apparently to keep cutting. I'm prepared to
arrest the entire 50-man camp if necessary."[72]

Outraged at the arrest of IWA members, union president Jack Munro
threatened a province-wide strike. Premier Bill Bennett asked Waterland to
end the dispute, but the number of arrests grew to 15 by the end of the
week. Although the area in question was small, the precedent-setting
nature of the conflict raised the stakes. Heavy rainfall and steep terrain on
the Charlottes meant that widespread application of Fisheries Department
regulations to prevent slides would remove a great deal of timber from
industry's grasp. The sides agreed to a temporary truce after the first week
as IWA, industry, provincial and federal representatives came together in
Vancouver to discuss a resolution. By this time Ottawa had new allies. The

United Fishermen and Allied Workers Union, Greenpeace, the Sierra Club and the B.C. Wildlife Federation all weighed in to support the salmon.[73]

The Riley Creek story soon assumed national significance, with NDP MP Tommy Douglas rising in the House of Commons to call for an end to the arrests. Leblanc blamed the province for going back on a prior agreement not to log Riley Creek. The combatants finally reached an accord on March 27, permitting Q.C. Timber to operate near the Riley Creek spawning grounds. Charges were dropped against the company and the loggers. The two levels of government agreed to more thorough consultation in logging plan approval, and initiated an $800,000 research effort into the habitat impact of steep-slope logging on the Queen Charlottes. For the United Fishermen and Allied Workers' Union and environmentalists, the deal represented a federal sell-out. Hart resigned his position as fisheries officer a few months later, expressing the same sentiments. That winter Riley Creek became the site of storm-induced slides, reinforcing the convictions of those who saw Waterland and the Forest Service as defenders of industry against the pressure of environmental values.[74]

In the aftermath of the slides, Waterland placed a moratorium on steep-slope logging on the west coast of the Charlottes. Q.C. Timber, now known as CIPA Industries, received approval of a five-year development plan in 1980, trading off rights to areas on Riley Creek for access to timber on Hangover Creek. A fisherman tried to revive the charges against CIPA at about the same time, but the provincial Crown prosecutor entered a stay of proceedings on the grounds of insufficient evidence. In a related case, the Supreme Court of Canada found the relevant clause of the Fisheries Act unconstitutional, ruling in 1980 that it violated the province's jurisdiction over forestry under the British North America Act. Forest-industry associations that had backed the appeal heaved a sigh of relief, but the events at Riley Creek continued to generate bitterness. When Peter Pearse conducted a federal fisheries inquiry in 1981, Northern Trollers Association spokesman John Broadhead said, the Riley Creek story "epitomizes everything wrong with our present techniques of enforcement. The law was deliberately broken, yet everyone involved managed to walk away from it." Pearse's reply that the Riley Creek issue merited study as "an example of what could go wrong" caused the forest industry some concern, and its own briefs emphasized the need for consultation between operators and agencies before matters reached the courts. Pearse's recommendations followed similar lines, and in placing habitat protection concerns in the context of over-fishing he gained praise from the Council of Forest Industries (COFI).[75]

Research on the interaction between forestry and fisheries on the Queen Charlottes and at Carnation Creek on Vancouver Island produced a set of 1988 logging guidelines, but only after COFI complaints about their cost and complexity delayed implementation. COFI described the new Coastal Fisheries/Forestry Guidelines as a commitment to translate "integrated resource management into reality", but the belated reintroduction of habitat protection after the demise of the Coast Logging Guidelines of the early 1970s drew skepticism from environmentalists, justified by a 1992 assessment that found industry compliance poor. Of 53 Vancouver Island streams examined, 34 were affected by debris or erosion. An appalled NDP Forests Minister Dan Miller promised a renewed emphasis on enforcement, penalties and stream rehabilitation by offenders. And federal Fisheries and Oceans Minister John Crosbie said, "This careless activity, close to fish-rearing and spawning habitat, by an industry that helped develop these guidelines, cannot go unchecked."[76]

Miller also promised to survey industry's performance on the Queen Charlotte Islands, but by 1992 logging on much of the archipelago had come to a halt as the result of an effective coalition forged by the Haida and environmentalists. During the 1970s concern over harvest levels on Rayonier's TFL No. 24 had grown, and the Riley Creek episode reinforced doubts about the Ministry of Forests' dedication to both sustained yield and integrated resource management. As Rayonier's TFL came up for renewal in May 1979, the Islands Protection Committee (IPC) voiced demands for a public hearing. Waterland, who had assigned the Environment and Land Use Committee (ELUC) Secretariat to study south Moresby Island while Frank Beban went ahead with logging Lyell Island, announced that he would grant Rayonier's application for a new licence when the existing contract expired. Multiple-use planning would guarantee the realization of both economic and wilderness values, he stressed, and Rayonier would be eligible for compensation if the withdrawal of productive forestland exceeded the legislated percentages.[77]

In pursuing an injunction against Waterland, IPC lawyers argued that since Rayonier was not the original licence holder, the TFL should be considered a new tenure. The Forest Act required public input in the award of new licences, and while the British Columbia Supreme Court denied that claim, a judge suggested that the petitioners should be consulted about the renewal. That led to meetings with senior Forest Service officials, gave the IPC a chance to examine agency files and permitted public review of management plans. Rayonier succeeded in achieving licence renewal shortly before selling out to Western Forest Products (WFP), but by late 1980, logging on the Queen Charlottes had become even more contentious. After disbanding its secretariat, the ELUC ordered the creation of a South Moresby Resource Planning Team coordinated by the Ministry of Forests. ELUC's evaluation of the team's four-option report began in 1983. WFP rejected all four options in favour of a plan that respected the firm's timber rights, involving withdrawal of no more than the five per cent of TFL No. 24's AAC permitted under the Forest Act without compensation.[78]

Not until 1981, when the Haida submitted their land claim to the federal government, did the drive for a wilderness area become explicitly bound up with the land-title issue. Haida policy statements took direct aim at the government-sanctioned rate of logging on the Queen Charlottes, TFL No. 24 management and the Ministry of Forests' disregard of the multiple-use concept. Over the next few years the forging of an alliance between environmentalists and the Haida led to consideration of a national park that would not infringe upon the comprehensive land claim.[79]

The dispute came to a boil in the autumn of 1985 when more than 70 people, mostly Haida, blocked Frank Beban's road on Lyell Island after the courts denied an injunction to halt logging. Arrests were made, with 9 Haida receiving suspended sentences, and suddenly the conflict attracted intense public interest. The province responded with more studies and considered affording park status to part of the island, but as the blockade continued more prosecutions followed. WFP offered to give up 70 square miles (181 square kilometres) of its TFL lands to resolve the dispute. By the end of 1985 Ottawa had entered the picture more directly, negotiating with the province over the national-park proposal. Waterland would soon exit the scene, forced to resign over his holdings in Western Pulp Partnerships, a corporate entity with a stake in Lyell Island timber. With the Nisga'a pressing their claim to the Nass River valley, and the Nuu-chah-nulth doing the same on western Vancouver Island in an effort to stop MacMillan Bloedel's logging of Meares Island, the Ministry of Forests was embattled on several fronts.[80]

The task of sorting out these and other land-use conflicts fell to the province's new Wilderness Advisory Committee, which recommended in March 1986 that WFP be allowed to log most of Lyell Island, except for a small ecological reserve, along with three other islands. In committee hearings, WFP warned of severe job losses and the high cost of compensation for lost timber given the lack of Crown forests available for a swap. President Roger Manning claimed that the province would lose 110 logging, 200 milling and 660 spin-off jobs. Federal studies indicated that greater revenues and employment would be achieved from tourism in the event of a national park, findings Manning termed "ludicrous" given the weather. Meanwhile, ongoing blockades on Lyell Island had brought the arrest of 72 protestors, witnessed by television cameras and journalists.[81]

By the spring of 1987 the proponents of a national park had clearly gained the momentum, and federal-provincial negotiations centred on the price Ottawa would pay for South Moresby. The announcement of a provincial moratorium on logging in the area that March produced outrage from industry, but gave additional time for discussions between representatives of the federal and provincial governments. Finally, on July 6, 1987, the haggling came to an end with the agreement that South Moresby would have a $106 million price tag for Canada. Of that, $32 million would go to building and operating the national park, $24 million would provide compensation to WFP and Beban, who suffered a fatal heart attack that year, and $50 million was reserved for regional tourism and small business development. The province would contribute another $8 million to the compensation package. WFP, which lost 15 per cent of its timber supply and saw its AAC drop from 432,000 to 130,000 cubic metres, claimed that the park "gutted a well-managed sustained yield forest".[82]

In the end Doman Industries, which acquired WFP, received $37 million in compensation for lost cutting rights. Loggers and companies questioned the benefits of Gwaii Haanas National Park, which went under joint Haida-federal management in 1993. In 1990 WFP executive Bill Dumont condemned the decision, pointing to the loss of 132 jobs at its camps and lower than anticipated tourist revenues. Others bemoan the decline of Sandspit, which lost about 200 residents by 1991. While the Haida welcomed the end of logging on the 1,470 square kilometres of Gwaii Haanas, they remained dedicated to the more significant resolution of their land claim. The federal government accepted the claim for consideration in 1983, but the province's refusal to recognize aboriginal title stalled negotiations. After withdrawal from the formal claims settlement process in 1989, the Haida monitored the Nisga'a negotiations and the Gitxsan-Wet'suwet'en Tribal Council's lawsuit as it wound its way through the courts.[83]

The Gitxsan had filed their claim in 1983, extending from Ootsa Lake to the headwaters of the Skeena and Nass rivers and from Burns Lake to Terrace, in response to the expansion of resource exploitation. While the *Delgamuukw* case inched along, the Gitxsan saw no alternative but to take direct action at the end of the decade when the Ministry of Forests awarded a new Forest Licence in the northern part of their claim and Westar prepared to begin logging north of the Babine River to feed its new Carnaby sawmill near Hazelton. Westar and a couple of Prince George groups competed for the new licence in 1988, disregarding the Gitxsan claim and raising concerns in Hazelton about its economic future should the timber go to Prince George for processing.[84]

When Westar built a bridge across the Skeena River about 40 kilometres west of Terrace, members of the Gitxsan Eagle Clan set up a blockade in February to protect their hereditary territory. Negotiations over a company offer to provide training and equipment for Gitxsan logging of the area followed, breaking down over the nature of Westar's 20-year logging plan. The Clan countered with a 250-year plan that would incorporate selective logging with hunting, trapping, tourism and protection of sensitive areas. With logging stalled in that area, tensions rose two weeks later when another blockade went up near Kispiox. As happened in other such incidents around the province during this period, truckers retaliated with their own blockade preventing Gitxsan people from leaving Kispiox. Relations among the Gitxsan became tense as well, due to the number of aboriginal loggers Westar employed in the upper Skeena valley.[85]

While these events unfolded, the Gitxsan-Wet'suwet'en Tribal Council went to court seeking an injunction against issuance of the new Forest Licence. They failed on that front, and Forests Minister Dave Parker awarded the licence to a Prince George consortium in May. The Tribal Council did succeed in obtaining an injunction against Westar's construction of a bridge across the Babine River, however, thanks to a ruling that resolution of the land claim should come first. When the Ministry of Forests showed no willingness to compromise elsewhere in the region, more roadblocks went up in the autumn of 1989, first at Kitwancool and then in the Suskwa and Kitwanga areas. On October 30, amid discussions that would eventually lead to the establishment of a regional First Nations-industry task force on forestry practices, blockades shut down Westar's activities in the entire Kispiox valley. Parker referred to the roadblocks as "outright lawlessness", but the government failed to secure an injunction against the series of stoppages. "Destruction is happening at such a rate that chiefs are concerned that while the court case is going on, the land will be destroyed," a Gitxsan official explained.[86]

The widespread use of herbicides to prevent brush encroachment on plantations provided a point of consensus among all residents of the Kispiox Forest District. When activists appealed every spraying permit in 1989, the Forests Ministry established an advisory committee that included representatives of the Gitxsan-Wet'suwet'en Office of Hereditary Chiefs, Suskwa Community Association, Kitimat-Stikine Regional Board, Kispiox Valley Community Association, Kitwanga Rod and Gun Club, Kispiox Valley Fishing Guides Association, Kispiox Valley Farmers Institute and Westar. In addition to considering the shortcomings of plantation silviculture in the region the committee drew attention to bleak timber supply estimates of the Kispiox TSA, which predicted the exhaustion of sawlogs in the Hazelton and Kitwanga areas within about 10 years.[87]

The Village of Hazelton drew all of the grievances that had festered over the 1980s together in a 1990 Forest Industry Charter of Rights, calling for a complete overhaul of provincial forest policy. Mayor Alice Maitland sent the Charter to municipal and regional district governments throughout British Columbia, arguing, "the owners of forest tenures have been shown to be motivated predominantly by profit considerations, and not by a fundamental concern for environmental and community stability." Functioning under the influence of its corporate clients, the Ministry of Forests clung to "management ethics of a bygone era". Disregard of legitimate First Nations land claims, poisoning of water supplies and massive clearcuts in every accessible valley demanded the introduction of a holistic, community-based approach capable of ensuring ecological diversity and a greater flow of economic benefits to resource-producing regions.[88]

Maitland had seen the industry around Hazelton evolve from a diverse structure of small, locally-owned mills to one dominated by Westar, which had recently laid off more than 100 workers after completing the computerized Carnaby sawmill. Hazelton's dependence on that plant, uncertainty caused by the Gitxsan-Wet'suwet'en land claim, loss of the Sustut Forest Licence to Prince George interests and a pervasive atmosphere of community insecurity gave rise to the Charter. The document demanded meaningful public input through citizen-controlled advisory/audit committees in each TSA, completion of a biophysical inventory prior to harvesting in watersheds, and a requirement that 75 per cent of the annual cut in any TSA be processed in that unit. "No corporation should be allowed to impoverish a community only for the sake of an increment of profit," the document read. Other proposed reforms included limitations on clearcut size, adequate leave strips along water courses, more local nursery production, industry-funded retraining for laid-off workers, and a rebate of forestry revenues to local and regional governments.[89]

The Hazelton manifesto's appeal for an immediate resolution of the aboriginal title question was denied a few months later, when Chief Justice Allan McEachern rendered his decision on the *Delgamuukw* case on March 8, 1991. Ruling against the Gitxsan and Wet'suwet'en, McEachern rejected any claim to aboriginal rights to land. After filing an appeal, the Gitxsan resorted once again to direct action with a January 1992 blockade of a Kitwanga Logging Company road. The owners secured an injunction in early February, but the Eagle Clan ignored the order to disperse. The blockade finally came down in return for a Ministry of Forests offer to discuss grievances, accompanied by the company's agreement to halt logging during negotiations, but the episode heightened tensions between non-aboriginal and aboriginal residents.[90]

The fall of 1993 brought yet another blockade, this time to halt logging development by the Rustad Brothers Company of Prince George. The Gitxsan-Wet'suwet'en case had entered its final stages, and despite business opposition to any recognition of aboriginal title, Mike Harcourt's NDP government moved to open treaty negotiations with the province's First Nations. The 1993 Court of Appeal decision provided further incentive, overturning a key element of McEachern's judgement. There had been no "blanket extinguishment" of aboriginal title with the establishment of British sovereignty, the court ruled, and non-treaty First Nations possessed a "limited form" of title, "other than the right of ownership", to some lands.[91]

Denied victory on the specifics of their claim, the Gitxsan-Wet'suwet'en appealed to the Supreme Court of Canada while cooperating with local conservation interests on land-use issues and entering treaty negotiations in 1994. The province dropped out in 1996 to pursue the *Delgamuukw* case, which ended in a December 1997 Supreme Court decision legitimating the concept of aboriginal title. While the court did not rule on the specific issues of the Gitxsan-Wet'suwet'en claim, the principles it set out involved a limited defence of the property rights attached to aboriginal title. These included not only rights to particular lands, but the resources on these lands. Continuous occupation after the introduction of British sovereignty in 1846 had to be proven, but even in this instance title did not confirm absolute ownership. The federal and provincial governments possessed the right to infringe upon aboriginal title in pursuit of legitimate objectives, but had an obligaticn to compensate First Nations for resource extraction on their territories, and a duty to consult before making major land-use decisions.[92]

First Nations greeted the *Delgamuukw* decision enthusiastically. Mel Smith, former constitutional advisor to the province, feared that the ruling undermined Crown ownership of 94 per cent of British Columbia, threw up insurmountable barriers to resource management and jeopardized private property. But *Delgamuukw* did not "confirm the actual existence of aboriginal title on any particular area or parcel of land", Gurston Dacks notes, and governments did nct change their negotiating position in any fundamental way. The Supreme Court's decision that aboriginal oral testimony and traditions merited the same consideration as documentary evidence greatly strengthened their position in the courts, but neither the legal route nor negotiations provided a speedy resolution of land-claims disputes. Mounting First Nations frustration over ongoing resource exploitation on traditional territories, expressed in the Gitxsan blockades, gradually fostered a willingness to negotiate recent "interim measures" agreements. The provincial government has also shown a grudging acceptance of its duty to consult with First Nations over land-use issues, although not without court challenges by the Haida to the recent transfer of TFL No. 39.[93]

For the Nisga'a, pursuing their claim outside the new treaty process, *Delgamuukw* lent validity to their oral history in negotiations. The question was, would anything be left of the Nass River valley when the talks concluded? After an initial series of announcements about large investments at its Prince Rupert and Terrace facilities, Repap's relationship to the region quickly turned sour. Rising to prominence in the mid 1980s, Repap's fortunes began falling about the time of the Westar acquisitions. Repap BC managed to open a state-of-the-art sawmill at Terrace, achieving a 20-percent reduction in the workforce while doubling the productivity of the existing plant. The firm also prcmised to spend $70 million modernizing Skeena Cellulose, but by early 1987 Repap itself was on the block. A merger with Avenor seemed to be in the works, but Avenor shareholders rejected the deal for the time being.[94]

A sharp recession coupled with bitter labour-management relations at Skeena Cellulose deepened Repap's troubles in the early 1990s. Losses at Prince Rupert topped $24 million in 1991 before pulp prices began to rebound. Repap's relationships with its contractors also suffered during the early 1990s, as truckers, logging operators and suppliers awaited payment for services. In 1995 the Terrace-based North West Loggers Association alerted NDP Forests Minister Andrew Petter to the serious effects delayed payments had on the region's economy, and warned of worse devastation to come should Repap "abandon its B.C. division and leave millions of dollars in unpaid bills to unsecured creditors". Meanwhile, Nisga'a leaders pleaded in vain for an interim measures agreement to halt logging while treaty talks proceeded. A 1993 report by Silva Ecosystems for the Nisga'a Tribal Council documented the existence of a 6000-hectare clearcut, failing plantations and extensive loss of fish and wildlife habitat on TFL No. 1.[95]

The Nisga'a called for an 84-per-cent reduction in the AAC of TFL No. 1, but despite a worsening reputation and poor financial performance Repap BC continued gobbling up operations and their timber rights. Acquisitions included Westar's Carnaby sawmill near Hazelton, Orenda Forest Products and a controlling interest in Buffalo Head Forest Products. These moves gave Repap an even more commanding position in the north-coast timber economy, its numerous Forest Licences and TFL providing an AAC of two million cubic metres. But the purchase of Orenda in 1996 came with strings attached, as the NDP government struggled to keep Skeena Cellulose afloat while retaining a measure of control over a company that employed about 1,450 mill workers and another 750 loggers. In a measure designed to maintain employment and community stability, Forests Minister Dave Zirnhelt set conditions requiring Repap to process Orenda's timber at Prince Rupert, continue providing work for its employees and contractors, work with the District of Stewart to establish a mill there, and complete plans for a $250 million modernization of the Skeena Cellulose pulp plant by June 30, 1997.[96]

Less than a year later, Repap BC's entire enterprise crumbled under the weight of debts amounting to $620 million. The immediate cause was a failed takeover of Repap Enterprises by Avenor in early March 1997. When Avenor shareholders rejected the inclusion of Repap BC in the deal, the parent company walked away from the failing operation, seeking court protection. The firm, now operating as Skeena Cellulose, sought protection under the Company Creditors Arrangement Act, but its two largest creditors withdrew their support. Owed $420 million, the Royal and Toronto Dominion banks refused to extend further credit until the provincial government and unions made concessions to lower long-term costs. The Ministry of Forests had already introduced a lower stumpage rate and deferred payments through 1997, but the Pulp, Paper and Woodworkers of Canada (PPWC) made clear its refusal to be "the scapegoat for the company's financial problems". Closure of the Prince Rupert mill and sawmills in Terrace, Hazelton and Smithers put 2,400 employees out of work that June, and affected thousands more. Contractors with unsecured debts of over $70 million also took a hard hit, and the impact rippled through the northwest economy as crews and suppliers went unpaid. "The Tree Farm Licence concept is supposed to help local communities and provide its communities with stability," Terrace contractor Ken Houlden exclaimed.[97]

In the wake of the closure of the northwest's largest employer, recrimination flowed freely. Repap BC officials faulted the NDP's high stumpage rates and Forest Practices Code for driving costs to unbearable levels. Terrace contractor Frank Cutler blamed Repap BC chairman George Petty for siphoning off money from the province to support the parent company's operations in Manitoba and Wisconsin. Others pointed the finger at the NDP for permitting Repap to expand without sufficient monitoring of the terms imposed in the Orenda acquisition. Still others, taking a more long-term perspective, argued that the Prince Rupert mill was a mistake from the outset, compounded by decades of Ministry of Forests efforts "to sustain the corporate entity, not to sustain forests and not to sustain communities". In a 1998 joint strategy for local control of forests, the Village of Hazelton and the Gitxsan-Wet'suwet'en Marketing Corporation interpreted the Skeena Cellulose debacle as the culmination of "a trend toward corporatism where companies end up setting policy for, and controlling, government".[98]

Whether the Skeena Cellulose problem presented the logical fulfilment of a flawed sustained-yield policy in the northwest or the recent machinations of a corporate shark, the operation's economic importance demanded a rescue mission. In late 1997, a last-minute compromise between the banks, the NDP and workers pulled it out of bankruptcy. The two banks agreed to write off $305 million in debts and invest another $94 million for plant modernization in exchange for a 55-per-cent interest in Skeena Cellulose. The province's commitment for a 25-per-cent interest came to $74 million. Local 4 of the Pulp, Paper, and Woodworkers Union took 20-per-cent ownership, swallowing a wage deferral package and the loss of 150 jobs.[99]

Premier Glen Clark defended the action as an investment in the northwest's economy, while critics called it a bailout to preserve Employment and Investment Minister Dan Miller's north-coast riding for the NDP. Either way, that left Skeena's contractors out in the cold. With no alternative but bankruptcy, they accepted 10 cents on every dollar Skeena owed them. But as industry analyst Jim Stirling noted, the story was not finished. Skeena's sawmills and extensive timber holdings were valuable commodities, but the pulp mill was a dinosaur, "an old and tired dinosaur". History might well repeat itself, he cautioned, with the pulp operation dragging down the entire company in future.[10]

Skeena Cellulose began turning out pulp again after a four-month shutdown, but before the end of 1997 the Clark government had to step in again. When the banks threatened to cut off funding unless they received $40 million in operating capital for 1998, the province acquired the Royal Bank's interest for $31 million, lifting its share to 52 per cent and a total investment of $240 million. Unable to find a private investor to take the operation over, the NDP and its partners operated Skeena Cellulose until the election of Gordon Campbell's Liberal government. But Skeena's problems did not go away. Stumpage charges along with logging and transportation costs made it more attractive for the company to barge chips from Alaska than rely entirely on its Crown timber supply. The cost to taxpayers reached $450 million as the entire northwestern timber economy struggled under the burden of American quotas on softwood lumber and tight Asian markets. By 2002 Skeena Cellulose was bankrupt again, and proved to be a headache for a new Liberal government bent on budget trimming.[101]

Desperate to extricate themselves from Skeena Cellulose, the Liberals succeeded in selling the Watson Island pulp mill, sawmills, and woodlands operation for $8 million in April 2002. The identity of the purchasers raised some eyebrows. Former Repap executives George Petty and Daniel Veniez, whose company had "precipitated a crisis and chain of events that left thousands of forest workers and suppliers out of jobs and millions of dollars owed" in 1996, were back with a holding company called NWBC Timber and Pulp Company. NWBC agreed to pay $2 million to Skeena's uninsured creditors in taking ownership, but ran into problems raising funds to activate the operational arm of the new entity, called New Skeena Forest Products. The City of Prince Rupert, owed $11 million in back taxes and facing the out-migration of residents, backed NWBC with a $1.5 million loan to facilitate the transaction. Prince Rupert residents even approved a deal that would permit the town to buy the pulp mill and TFL for $20 million and sell it back to NWBC for annual payments of $150,000. That, in theory, would help overcome NWBC's financing difficulties and revive operations by the community's major employer. The Terrace Town Council, on the other hand, opposed NWBC's purchase of Skeena Cellulose.[102]

NWBC President Veniez adopted an aggressive stance toward unions and contractors, a philosophy of taking "a chainsaw to what he sees as a culture of entitlement", one writer observes. He forced concessions in the form of "Fresh Start" agreements on PPWC Local 4 at Prince Rupert and the IWA at Terrace, terminated agreements with logging contractors, and offered timber to the Gitxsan in the Terrace area. Veniez stated that start-up of operations hinged on realization of the proposed $20 million agreement with the City of Prince Rupert, but some felt that he intended to break the region's woods unions, and others were offended by his export of raw logs. Whatever the outcome of the New Skeena venture, a resolution of the problems that beset the plant responsible for 25 per cent of Prince Rupert's economy and 20 per cent of its workforce was vital to the community's well-being in the short-term.[103]

Recently, however, patience has worn thin and suspicion deepened toward Veniez and Petty. While attributing the northwest's economic decline to a "regressive regulatory climate and pervasive, almost stifling entitlement culture", Veniez was unable to secure financing and this pushed back the timetable for a reopening. New Skeena was forced to file for bankruptcy protection in late 2003 after municipal governments in the region ordered bailiffs to seize equipment at the mills to cover unpaid property taxes. Veniez held out hope that Gordon Campbell's Liberal government would come through with help, but the action of the municipalities left him exasperated. "We've been a company coming out of a bankruptcy, for goodness sakes," he observed. "How are we going to attract investment to our heartland communities in British Columbia if companies who are distressed were obliged to pay punitive property taxes."[104]

For all of Prince Rupert's problems, the presence of pulp manufacturing there had the virtue of economic development based on local processing of resources. The Bella Coola valley had never enjoyed even the unequal rewards of such a policy, nor did the entire central coast after the Ocean Falls closure. Given the meagre benefits industrial forestry afforded, local officials continued to advocate a more diversified model of economic growth, one with a central place for wilderness tourism. But even that vision became more problematic as logging rates soared during the 1980s to feed outside mills. Rather than heed the Bella Coola Regional Study's proposal for local manufacturing of unalotted Crown timber, awards continued to go to major licensees such as Fletcher Challenge to supplement its old Crown Zellerbach tenures. Moreover, after merging the central coast PSYUs into the Mid-Coast TSA, the Ministry of Forests avoided reductions in the allowable cut by committing operators to harvest the poorer-quality timber at higher elevations. In fact, the annual harvest increased steadily over the decade, justified by the expectation of a balanced cut of the valley bottoms and hillsides.[105]

By the end of the 1980s the Central Coast Economic Development Commission deplored both the absence of local manufacturing and overcutting by the major licensees. The addition of helicopter logging sales in a larger AAC for the Small Business Enterprise Program in 1989 triggered a protest from Hagensborg Economic Development Officer Patricia McKim-Fletcher. Her agency had nothing against helicopter logging, but with an ongoing drain of resources at unsustainable levels. "What we are saying is that the Central Coast has been on the overcut for many years," McKim-Fletcher declared, "and most of the best wood is gone with little economic benefit to our communities."[106]

Fletcher-Challenge was the prime culprit in the Bella Coola area, consolidating operations there in a campaign of exploiting low-elevation Douglas-fir for processing on Vancouver Island and the Lower Mainland, and leaving more expensive stands on the slopes. The firm had exceeded its Mid-Coast AAC by 50 per cent during the 1980s, McKim-Fletcher charged in 1990, an accelerated cut that would bring the valley to the brink of falldown within 15 years while "destroying our tourist potential and degrading our environment". The answer was a Forest Licence for the Bella Coola valley, enabling the community to pursue real sustained-yield management, selective logging, economic diversification and revenue generation to meet regional needs. "We are only exploiting our local resource for the benefit of distant owners and shareholders," the proposal observed. "Changing economies at a local level within a context of sustainable diversity and economic self-sufficiency should be the ultimate goal for the residents of the central coast."[107]

Bella Coola.
BC Archives NA-26503

Ministry of Forests analysis of harvesting rates and practices on the Mid-Coast TSA between 1986 and 1989 lent official weight to this diagnosis, if not the region's prescription for a cure. Operators such as Fletcher Challenge, International Forest Products (Interfor), MacMillan Bloedel, Mayo Forest Products and Doman Industries had failed to re-direct harvesting into the low-quality stands, the condition for maintaining and increasing cutting rates during the 1980s. This, along with increasing pressure for wildlife, fisheries and recreational resources, mandated a 34-percent reduction in the Mid-Coast TSA's annual harvest in 1992. In announcing the cutback Chief Forester John Cuthbert promised closer Forest Service monitoring of utilization practices, but the long-term outlook was undeniably bleak. Within 50 years the Mid-Coast district's cut would have to come down from nearly 1,000,000 to 550,000 cubic metres to achieve sustainability, the ministry's 1994 Timber Supply Review concluded.[108]

Facing similar reductions elsewhere along the coast, except in the far north, a recession-plagued industry asserted its vulnerability to market pressures, the demands of environmentalists and First Nations, and the high cost of harvesting increasingly remote Crown timber. Mid-coast operators argued that the ministry's supply analysis underestimated the productivity increases that could be expected from managed second-growth stands, and called for incentives to harvest low-quality sites. The Central Coast Economic Development Commission and Central Coast Regional District countered by attributing the region's high unemployment rate to the operators' lack of initiative during the 1980s. All interests pleaded for a gradual implementation of AAC reductions, for industry to preserve its access to the resource and for local parties to minimize community impacts.[109]

For Mike Harcourt's NDP government, elected in October 1991, the environmental movement's growing international prominence and alliances with First Nations presented a significant challenge in balancing its "red" and "green" constituencies. After the relatively painless decision to award Class A park status to the Khutzeymateen, pressure for preservation of the Kitlope River valley southeast of Kitimat ascended to the top of the preservationist agenda. After buying B.C. Timber's Skeena Lumber Division sawmill at Terrace in 1983, West Fraser and Eurocan had made preparations to log the Kitlope watershed portion of TFL No. 41, and by 1987 were involved in discussions with the Haisla people who had filed a land claim over the area.[110]

The involvement of Ecotrust, a Portland-based environmental organization, further complicated harvest planning. In 1990 the Earthlife Canada Foundation and Conservation International funded a study by Keith Moore of the remaining intact coastal watersheds. Moore identified the Kitlope watershed as the largest on the British Columbian coast, and Conservation International began focusing public attention on the region's special, pristine features. Spencer Beebe and Ken Margolis left the organization to found Ecotrust, and went on to play a key role in preserving the Kitlope. In the spring of 1991, the Social Credit government announced a decision to defer logging in the area for two years. West Fraser responded that the area was anything but unique, and cautioned that the Kitlope represented 12 per cent of the Terrace mill's future annual wood supply. Since the firm planned to log only 6 per cent of the valley, however, the deferral did not provoke immediate outrage. Harcourt's new NDP government then extended the moratorium and considered the Kitlope for inclusion under their Protected Area Strategy, which aimed to set aside 12 per cent of the province's land base.[111]

Shortly before buying out its Finnish partners in Eurocan in 1993, West Fraser went to the Haisla with a deal, offering to preserve 100,000 hectares of TFL No. 41, about one-quarter of the valley. Logging of the remaining 300,000 hectares would provide employment for about 65 Haisla, with their representatives joining a committee responsible for setting the AAC. The Haisla rejected this "jobs for consent to log" arrangement, affirming their opposition to any logging of the watershed. Rather than taking to the courts, the Haisla worked closely with Ecotrust in a campaign to highlight the need for preservation of the Kitlope's cultural and aesthetic values along with its fisheries and wildlife. In addition to supplying the Haisla with scientific data, Ecotrust established a local conservation body at Kitimat called the Nanakila Institute, questioned the economic viability of Kitlope logging and asserted the valley's wilderness-tourism potential.[112]

By the end of 1993, support for co-management of the area by the Haisla and provincial government took the spotlight. The Valhalla Wilderness Society, Sierra Club and Western Canada Wilderness Committee all promoted protection, but it was Ecotrust's strategy of working closely with the Haisla that proved decisive. Ecotrust's Ken Margolis called the Kitlope campaign a "cakewalk" compared to the struggle to save Clayoquot Sound from logging, given the relatively low number of jobs and timber values at stake. "A quick decision to·protect this watershed seems in order," journalist Craig Orr concluded. West Fraser's Hank Ketchum would eventually come around to this point of view, but not without considerable opposition from woodlands managers. Northwest Timber Operations manager Bruce MacNicol challenged the "biased opinions, inaccurate information and misconceptions" circulating around the Kitlope early in 1994. Modern technologies not only made the logging option sound economics, the company's plan to log just 50,000 cubic metres annually would impact only five per cent of the total area over the next 120 years. Whatever the outcome of the Haisla land claim, in the interim the Kitlope logging plan would benefit them, meet corporate wood supply objectives and satisfy the environmental community.[113]

It came as something of a shock, then, when Mike Harcourt announced on August 16, 1994, that West Fraser had relinquished its Kitlope cutting rights in TFL No. 41 without compensation, and the entire 317,000 hectare watershed would be closed to industrial forestry under a plan for joint management by Victoria and the Haisla. Despite resistance from within industry ranks and his own company, Hank Ketcham earned praise for extricating West Fraser from the controversy. In so doing, he also gained his company a "green" reputation among customers in the public relations wars over British Columbia forest practices. With another provincial Timber Supply Review looming and further downward adjustments in harvest rates inevitable, industry predicted that the Kitlope result would only exacerbate the timber shortage.[114]

Indeed, by the end of 1994 the NDP's protected area policies, land-claims negotiations and Forest Practices Code convinced established economic interests that government interventions would decimate the northwest's economy. All of this, in conjunction with cancellation of the Kemano Completion Project, creation of the 946,000-hectare Tatshenshini-Alsek Wilderness Park and Nisga'a land-claim negotiations were interpreted as elements of an economic doomsday scenario. "The area's economy is slowly being strangled as it is cut off from resources," declared the pro-business *BC Report.* "Unless we get new mines and make wood available to at least feed existing mills, our communities are going to enter a pattern of decline," Stewart Mayor Andy Burton predicted early in 1995.[115]

Developments over the next year or two did nothing to lift spirits. Analysis of the Queen Charlotte TSA indicated that the 515,000-cubic-metre annual cut would have to be slashed by 14 per cent immediately, followed by 12 per cent reductions each decade for the next six decades before harvests could gradually rise to a sustainable level of 248,000 cubic metres. Adding marginal stands to the inventory along with more optimistic growth projections enabled Chief Forester Larry Pedersen to keep the reduction to 7.5 per cent in 1996. But that August, Greenpeace and Haida demonstrators blocked the departure of a MacMillan Bloedel barge to protest the unsustainable harvest rate and export of logs to southern mills. The province responded in September by signing an agreement with the Islands Community Stability Initiative to provide community forest tenures and accept the input of an Islands Forest Council in planning.[116]

But by this time the Ministry of Forests had even more significant potential problems involving a 1995 Haida challenge to the renewal of MacMillan Bloedel's TFL No. 39 on the grounds that they had not been consulted. After losing in the BC Supreme Court, the Haida achieved something of value in a late-1997 Court of Appeals decision. Although the renewal went through, the court recognized that their unextinguished title to the land might include a legal right to the forests. The Haida considered this a significant step in achieving their claim to what they called Haida Gwaii, as well as leverage to negotiate interim agreements to protect their rights to lands and forests.[117]

The Council of the Haida Nation was back in court again in 2000, challenging MacMillan Bloedel's sale of TFL No. 39 to Weyerhaeuser on the same grounds. That prompted the province to withdraw from discussions of interim-measures agreements during treaty negotiations. In a 2002 ruling the BC Court of Appeals held that the Crown and Weyerhaeuser had failed in their duty to consult and find a "workable accommodation" with the Haida, violating the principles of the 1997 *Delgamuukw* decision in the transfer of TFL No. 39. The matter went on to the Supreme Court of Canada while the Haida filed suit claiming title to the Queen Charlotte Islands, but was clear that in the "new legal reality", government and industry could no longer disregard First Nations interests in forests.[118]

As the pressures on accessible timber supplies mounted, the Ministry of Forests and companies looked north to the Cassiar Forest District, 150,000 square kilometres of fragile forestland stretching from the Stikine River north to the Yukon border. The province's last forest frontier had no large-scale licensees in 1994, and the few sawmills cut only to meet local demand, but the southern portion of Cassiar gained importance to a "fibre-starved industry" despite its remote location. With the Kalum TSA just to the south due for an AAC reduction to prevent future timber shortages, the ministry concluded in 1995 that the Cassiar District could sustain a ten-fold increase, from its current meagre cut of 87,000 cubic metres to 842,000. Environmental organizations raised the alarm, attributing the push north to the desperate scramble for wood by mills in Terrace, Houston, Smithers and Quesnel. An all-out assault on the scrub forests of the Cassiar would "imply despoiling an enormous amount of pristine land", environmentalists charged, extending ecological "pillage" carried out by Buffalo Head and Orenda in the Kalum District.[119]

Trappers and guides joined in warning about the difficulty of renewing forests in the Cassiar region's ecosystem, and in questioning the wisdom of harming wilderness-dependent businesses to create a few logging jobs. The David Suzuki Foundation's Jim Fulton said that clearcutting these boreal and sub-boreal forests would "doom the Cassiar to looking like the dark side of the moon". Other opponents, referring to the 1980s experience of log-truck caravans bound for Terrace, Smithers or Prince George, doubted that a higher cut would bring any new processing jobs to Stewart. Chief Forester Pedersen announced a new AAC of 400,000 cubic metres in November 1995, up from 140,000, but NDP Forests Minister Andrew Petter declined to accept the recommendation. Instead, he set the AAC at half the amount announced by Pedersen, pending the development of a land-use plan for the region. Both the Kalum and Nass TSAs underwent minor reductions at the same time, amid expectations of further withdrawals to come as the Nisga'a treaty negotiations neared a conclusion.[120]

The Nisga'a Tribal Council and the federal and provincial governments reached an agreement-in-principle on the treaty in February 1996, providing for a cash payment of $190 million and Nisga'a government ownership of nearly 2,000 square kilometres of land. The Nisga'a assumed responsibility for about 45,000 hectares of productive forestland in the North Coast, Nass and Kalum TSAs, and for TFL No. 1. The province would no longer collect stumpage fees or rents on these lands, revenues that would now go to the Nisga'a, in exchange for reduced provincial and federal transfer payments. A socio-economic study indicated that "possible reduction or transfer of activity ... will be minor relative to the size of the regional industry," and the treaty would include a transition period to minimize non-Nisga'a job losses.[121]

Despite such assurances, industry raised several concerns with the province's Select Standing Committee on Aboriginal Affairs. The Truck Loggers Association's policy statement insisted that the negotiations produce "certainty, stability, finality and definition of rights for all parties [while] avoiding disruption of existing stakeholder rights of interests". Crown lands covered in forest tenures should not be included in a settlement without the holder's agreement, but in the case of unavoidable transfers, compensation must be paid to licensees, contractors and workers. Overall, the Truck Loggers concluded that the agreement-in-principle's vague provisions threatened "the health of the B.C. forest industry and the economic well-being of the province".[122]

These same concerns surfaced when the parties signed the Nisga'a Final Agreement in 1998. The Council of Forest Industries objected that the agreement would expropriate licensees' holdings without a guarantee of compensation, suggesting that loss of production or mill closures would prompt legal action if investments had been made as a condition of the tenures. The Final Agreement attempted to address these concerns with the inclusion of a five-year transition period during which licensees could continue harvesting on Nisga'a lands. Stumpage and licence fees collected by the province during this period would be reimbursed to the Nisga'a, who would take full control over forest management and revenue collection in 2005. While able to introduce their own management standards at that time, a clause dictated that these meet or exceed provincial requirements. Another provision required the Nisga'a to maintain the wood supply to local mills to avoid disruption of operations during the transition period. Nor could the Nisga'a establish a processing facility for 10 years after the final agreement.[123]

Almost 75,000 cubic metres of wood was harvested on the transition licences in 2001 and 2002, under terms requiring licensees to allocate half their logging to Nisga'a contractors. The Nisga'a Lisims government intends to build a sawmill capable of providing 80 or more jobs after the 10-year restriction expires, but in the meantime the Laxgalt'sap Forest Corporation employs 30 Nisga'a people, and their contractors are doing most of the logging. Skeena Cellulose received compensation for the portion of TFL No. 1 transferred, and in 1998 the Truck Loggers Association signed an agreement with the province providing for compensation in the form of opportunities on Crown land in the event of members losing work due to the Nisga'a Final Agreement.[124]

The Ministry of Forests announced AAC reductions for the Kalum, North Coast and Nass TSAs and TFL No. 1 when the Nisga'a Treaty came into effect in 2000, reflecting their reduced land base. But in 2003 First Nations controlled only about 3 per cent of the province's forest tenure. The forestry revitalization plan announced that year by Gordon Campbell's Liberal government withdrew timber from long-term tenures for redistribution to First Nations, communities and woodlot owners, building upon previous legislation that enabled aboriginal groups to apply for tenures without compensation. Seven agreements had been signed by June 2003, involving First Nations rights to 1.1 million cubic metres of wood.[125]

No one knows precisely what the Liberal measures will mean, and First Nations are by no means united in their praise. Much clearer, perhaps, is the history of legal wrangling and militant direct action that gave government and industry an incentive to forge less disruptive relations with First Nations. As a Council of Forest Industries representative put it recently, "industry members are anxious to see the treaty process make progress, and to try and improve the operating and investment climate." A good deal of this motivation is derived from the fight for the forests of the central coast during the 1990s, involving what we know now as the Great Bear Rainforest. The relationship between environmentalists, workers, First

Nations and companies in this campaign defies easy generalization, but consideration of the events provides additional insight into current developments.[126]

With no victories to claim on the mainland coast after the Kitlope, a number of environmental organizations began developing more ambitious strategies for the region during the mid 1990s. Peter McAllister of the Raincoast Conservation Society (RCS) called for a conservation campaign after an up-coast expedition in the summer of 1993. Clearcutting by Doman Industries, Interfor and Western Forest Products (WFP) would continue to decimate lands claimed by the Heiltsuk at Bella Bella unless a Clayoquot-style protest sparked public outrage. And while the economic needs of the Heiltsuk would likely see them engage in some logging, the RCS saw their stewardship "as the environment's best bet".[127]

By the summer of 1995 the RCS, Sierra Club of BC, BC Wild and Valhalla Wilderness Society were working on a "Spirit Bear Wilderness" proposal for Princess Royal Island and parts of the adjacent mainland, held under licence by WFP. Wayne McCrory, the wildlife biologist whose efforts had been pivotal in the Khutzeymateen receiving protection as a Grizzly Bear sanctuary, now proposed park status for Princess Royal Island, Swindle Island and several mainland inlets to preserve the habitat of the white Kermode Bear (a rare colour form of the Black Bear). The Forest Action Network (FAN) opened an office at Bella Coola that summer to work with the Nuxalk First Nation in protecting King Island from logging by Interfor. In September FAN and Nuxalk activists set up a blockade on the island, leading to several arrests after Interfor obtained a court injunction against the protest. Meanwhile, McCrory's Valhalla Wilderness Society and the Great Bear Foundation worked with the Kitasoo First Nation, natural allies in the campaign to stop WFP activities within the proposed sanctuary.[128]

In July 1996, after having mixed success with its Commission on Resources and Environment land-use planning effort in other parts of the province, the NDP government announced its intentions to develop a Land and Resource Management Plan (LRMP) for the central coast. The LRMP brought civil servants and over 40 stakeholder groups to the planning table to create protected areas and sustainable resource management systems for the central coast, with linkages to the First Nations treaty process. The Sierra Club established an office in Bella Coola to participate more fully with local interests in the LRMP, but by this time FAN had worn out its welcome.[129]

At a community forum that August, the Sierra Club took pains to distinguish itself from FAN and its direct-action tactics. Nuxalk Chief and Owikeeno/Kitasoo/Nuxalk Tribal Council member Archie Pootlass said that the number of families in his band drawing support from the forest industry had dwindled from 65 to 9. A majority of Nuxalk had voted to evict FAN from the reserve, but when asked directly, a FAN representative

White Kermode bear on Princess Royal Island, 1920s.
BC Archives D-03616

declined to offer to leave. Why not blockade the Legislature in Victoria instead of local workers, an audience member asked? "The government has got our people just where they want us: split down the middle," said Jack Edgar. "FAN will be here and then gone, Sierra Club will be here and then gone, but I'm going to be living here for the rest of my life." But if consensus on ancient-forest logging proved elusive at Bella Coola, most could agree that mid-coast communities should derive more financial and employment benefits from forestry and exert greater influence over land-use decisions.[130]

Establishment of the central-coast LRMP and a 1995 Grizzly Bear Conservation Strategy failed to pacify environmentalists. While the former ground toward a conclusion, clearcutting would continue, and the latter would not be permitted to interfere with more than one per cent of the central coast's AAC. And the Forest Practices Code would merely disperse clearcuts over a wider area. Peter McAllister charged, allowing companies "to invade remaining coastal watersheds at an accelerated pace". An LRMP process for the north coast would not even begin for two years, giving industry plenty of time to finalize harvest plans for areas such as the prized Ecstall River valley. Determined to prevent what they considered Ministry of Forests liquidation of the "Great Bear Rainforest" 14 environmental groups organized the Coastal Rainforest Network in June 1996. A couple of months later Greenpeace demanded a complete halt to logging in British Columbia's old-growth forests.[131]

By early 1997 the Kermode bear was on its way to becoming an international symbol of the campaign, setting the stage for a conflict-filled summer. FAN organized a European tour with Nuxalk representatives, and that April Greenpeace initiated a boycott campaign among the customers of

British Columbia forest companies. When Greenpeace participated with FAN and Nuxalk activists in a blockade of Interfor operations at King Island, and WFP had to obtain an injunction against protestors at Roderick Island, government officials urged the organization to work with First Nations and other stakeholders in the LRMP process. "It is regrettable that Greenpeace has chosen to go the route of conflict and confrontation," NDP Forests Minister Dave Zirnhelt said. "The bottom line is that they are breaking the law and their activities are not welcome – not by Western Forest Products, not by the Kitasoo First Nation, and not by British Columbians who believe in responsible solutions to land-use issues." According to a *Truck Logger* story, the Kitasoo had told Greenpeace's Tzeporah Berman to move on when the organization arrived, and later issued a news release denying that they had "invited foreign environmental groups into the area to protect either bears or any other aspect of the environment". Nuxalk Chief Archie Pootlass would also make clear that the two Nuxalk protestors at King Island did not represent the interests of his people. "Your actions are driving a wedge between families in the Nuxalk Nation," he told Greenpeace. Premier Glen Clark used stronger rhetoric toward Greenpeace and its supporters: "Environmentalists who choose to work with American interests against our industry and jobs are enemies of British Columbia."[132]

If relations between environmentalists and First Nations had grown increasingly complex, those rooted in class tensions came to the surface along the Vancouver waterfront between June 28 and July 4, 1997. The King Island and Roderick Island protests had deprived IWA-Canada Local 1-71 members of work, and the union faced significant job losses depending on the outcome of the Great Bear Rainforest campaign. The union sued the Roderick Island and King Island protestors for lost wages, but the new president, Dave Haggard, also resorted to direct action when the Greenpeace ships *Arctic Sunrise* and *Moby Dick* docked at Vancouver to pick up supplies on June 28 to support the mid-coast initiative. An IWA information picket line went up immediately, and a few days later union members encircled both with boomsticks. Unable to depart, Greenpeace officials asked for dialogue. "Greenpeace has had more than its fair share of chances to join in legitimate dialogue over land-use ... during the last several years," Haggard told the media. "They are a bunch of hypocrites who set up illegal blockades preventing our members from going to work, and now that they are getting a taste of their own medicine they don't like it one bit." After a week Vancouver Port Police removed the booms, allowing the ships to escape, but the event and a similar confrontation at Squamish that summer led some to detect a shift in public sentiment against environmental organizations.[133]

Perhaps the most significant outcome of the 1997 protests involved a Protocol Agreement developed by central-coast First Nations that July, asserting their jurisdiction over the lands, waters, air and natural resources "given to us by the Creator" and sustained for thousands of years. The protocol did little immediate good on King Island when Chief Pootlass presented it to environmentalists. During an ensuing debate about the need for respect of aboriginal sovereignty they expressed attitudes both "crude and totally disrespectful", Pootlass claimed, according to the *Truck Logger*'s account. That disappointment could not obscure the more important fact of First Nations speaking with one voice against industry, government and environmental groups.[134]

Industry was busy as well as the Central Coast LRMP process got underway. Interfor, with the largest holdings at stake, approached the RCS and three other groups in 1998 with a proposal for a temporary truce. In the end, Interfor promised to continue logging in thirteen areas in exchange for freedom to log in three others, while the groups participated in the LRMP process. But Greenpeace continued to campaign for a European and American boycott of products from the Great Bear Rainforest and other old-growth forests, ultimately succeeding with customers such as Ikea and Home Depot. In 1998 MacMillan Bloedel responded by announcing a gradual phasing out of clearcutting in favour of variable-retention logging. West Fraser took a more drastic step in 2000, selling its north-coast Forest Licence to a contractor. The firm retained its Prince Rupert sawmill and the Kitimat pulp mill, but attributed its withdrawal from north-coast logging to harvesting difficulties and the need for a separate marketing strategy. The Ketcham family and West Fraser had given up the Kitlope without compensation as a legacy, Vice-President Wayne Clogg said, "but there was very little recognition from the environmental community and no recognition from the group now campaigning on the coast [Greenpeace]."[135]

Given the acrimony and distrust of the 1990s, it came as something of a surprise when the Coast Forest Conservation Initiative (CFCI) came to light in March 2000. CFCI involved an unlikely partnership of environmental organizations and forest companies whose Joint Solutions Project would attempt to bypass the Central Coast LRMP, bogged down as stakeholders struggled to reach consensus. Six companies: Canadian Forest Products, Interfor, WFP, Fletcher Challenge, West Fraser and Weyerhaeuser, recent purchasers of MacMillan Bloedel, formed the CFCI in January 2000 in response to the Great Bear Rainforest boycott campaign, seeking a working relationship with the environmental community. They approached and gained commitments from Greenpeace, the Rainforest Action Network, ForestEthics and the Sierra Club of BC to collaborate in finding a long-term peace in the war of the woods.[136]

Like the earlier agreement forged by Interfor, Joint Solutions called for a "conflict-free period" during which the companies would leave untouched up to 140 places for 18 months while the environmental groups curtailed their boycott effort. Discussions would follow, aimed at a permanent resolution of coastal land-use conflicts. Neither the province nor the woods unions were participants, however, and the CFCI began to disintegrate. Interfor quickly dropped out, along with West Fraser after the sale of its coastal Forest Licence. Interfor CEO Duncan Davies attributed the firm's withdrawal to a flawed process that had insufficient regard for communities, First Nations, contractors and workers, but Interfor also had the most to lose. Half of its AAC originated from the areas in dispute, while Canadian Forest Products had none at all and Weyerhaeuser only a small portion of its cutting rights at stake. The provincial government was also unhappy with the CFCI; even though its own LRMP had broken down it could not support a parallel planning process that did not include First Nations and other stakeholders. The Truck Loggers Association provided yet another very important source of opposition. Chairman Jack McKay described the CFCI as "an ill-advised effort to placate some large environmental groups … by shutting down logging in major areas of the central and north coast."[137]

By the summer of 2000 the CFCI, under attack on several fronts, took a hiatus in order to broaden its base of support. Meetings with government officials apparently followed, after which the remaining companies and Norske Canada agreed with the environmental groups to develop an "ecosystem-based model for conservation and management of coastal forests that fully integrates social, economic and ecological needs". This would involve First Nations and other interests along the central and north coast and Haida Gwaii, and coordinate with the LRMP process. The Truck Loggers Association immediately attacked the revived CFCI as an attempt to undermine the province's sovereignty. The advertised shift to a downsized ecosystem-based industry was a response to the "feverish prescription of foreign eco-bureaucrats," wrote Anthony Toth, "and contrary to the wishes of the B.C. Government, communities, workers and First Nations people who actually live, work and vote here." The LRMP process should go forward, he urged, without the interference of those who sought to shape its outcome to meet their own needs.[138]

With the Central Coast LRMP not yet complete, the AAC announced for the Mid Coast TSA in March 2000 maintained current harvest levels. The LRMP's outcome would be taken into account in future reviews, Chief Forester Larry Pederson explained. Another initiative had appeared by this time as well. That March the David Suzuki Foundation hosted the Turning Point conference, bringing First Nations leaders from the central and north coast and Haida Gwaii together to discuss issues of resource management and community sustainability. That culminated in the Declaration of the First Nations of the North Pacific Coast, reflecting concerns that the drawn-out processes of treaty negotiation and land-use planning would permit depletion of land and marine resources. The declaration set out the interests shared by the Council of the Haida Nation, Gitga'at First Nation, Haisla Nation, Heiltsuk First Nation, Kitasoo/Xai'xais First Nation and Metlakatla First Nation, and the Suzuki Foundation undertook to produce a set of sustainable forest management principles on their territories.[139]

Finally, on April 4, 2001, the provincial government announced its Interim Land Use Plan for the Central Coast along with a Protocol Agreement on Interim Economic Measures with area First Nations. The CFCI took credit as a catalyst in the achievement of what came to be known as the Great Bear Rainforest Agreement. The province, environmental groups, companies, First Nations and communities, signed on to what Premier Ujjal Dosanjh called a "hard-won consensus aimed at saving areas of global significance".[140]

But the accord was not set in concrete. Twenty large valleys of some 650,000 hectares would be protected, with logging deferred for up to two years on another 68 valleys covering 880,000 hectares, pending finalization of the Central Coast LRMP. Ecosystem-based management would be followed in order to achieve Forest Stewardship Council eco-certification of products, with advice from an independent team of scientists incorporating the traditional ecological knowledge of First Nations. "International markets want resolve on issues involving critical ecosystems and endangered forests on the B.C. coast," said a Weyerhaeuser vice-president. The CFCI's environmental partners also gave an enthusiastic endorsement. A Sierra Club official called the deal "a real turning point for the future of B.C.'s rainforests". The market campaign would be suspended while ForestEthics worked with stakeholders to achieve a final resolution, Tzeporah Berman declared. The Forest Action Network, which did not sign the accord, took a more reserved attitude. Would planning follow a true ecosystem model? Were the protected areas sufficient to ensure biological diversity?[141]

Workers and contractors, facing a 15-per-cent reduction in the AAC, also had reservations. To bring IWA-Canada and the Truck Loggers Association on board, the NDP government appointed a negotiator to develop an economic package intended to soften the impact of central coast land-use decisions. The government committed $10 million through Forest Renewal BC, the agency the NDP had established in 1994 to assist workers impacted by the creation of protected areas. Funded with the revenues generated by higher stumpage fees, FRBC invested in reforestation, silviculture, environmental restoration and research during the late 1990s. But not until government doubled its financial commitment did the IWA and Truck Loggers Association give their formal assent to the central-coast land-use plan. The major companies were to contribute matching funds for short-term support of those affected by the withdrawal of working forests.[142]

The First Nations protocol on interim measures and planning drew on the Turning Point Initiative, giving the Haida (Old Masset and Skidegate Councils), Gitga'at (Hartley Bay), Haisla (Kitimat Valley), Heiltsuk (Bella Bella), Kitasoo/Xai'xais (Klemtu) and Metlakatla First Nations a new role in decision-making and management. Through it the province would work with them in defining principles and anticipating the results of planning processes, and consider land-use plans submitted by the First Nations. In the event that they could not agree with the recommendation of a planning forum, the province agreed to the establishment of a "government-to-government process" to resolve the issue. The interim measures component obligated the province to facilitating First Nations involvement in forestry through joint ventures, the provision of tenures, and employment in silviculture.

Although the Protocol Agreement did not apply to many First Nations involved in the Central Coast LRMP process, its inclusion in the Great Bear Rainforest pact drew applause. "I am proud that the B.C. government has signed this agreement," David Suzuki said. "This would lead to a sustainable, diversified economy for generations to come." But as the Central Coast LRMP talks continued and the right-wing Liberals came to power environmentalist enthusiasm waned somewhat. Budget cuts, layoffs, splitting of the Ministry of Environment into two separate entities, and a lifting of a moratorium on Grizzly Bear hunting reflected a harsher neoconservative attitude to environmental protection. Still, environmentalists such as Wayne McCrory were encouraged by the May 2002 announcement that the 20 key areas of Great Bear Rainforest selected for protection in 2001 would be given formal status, even if a decision on other "deferred" areas awaited further study.[143]

By the beginning of 2003, however, with the moratorium on logging in the disputed areas set to expire in a few months, the era of good feelings among government, environmentalists, First Nations and the forest industry seemed at an end. In January the RCS, Suzuki Foundation and Forest Watch released an audit of every central-coast cutblock logged since April 2001. While no one had developed a precise definition of ecosystem-based management in the Great Bear Rainforest, the organizations agreed that large-scale clearcuts and logging the banks of fish-bearing streams would violate any reasonable standard. The audit revealed that, on 72 per cent of the 227 logging sites, 80 per cent or more of the trees had been removed and only 4 per cent of the plans provided for buffers along small streams.

The response to what Suzuki called the "business as usual" practices of the companies and Ministry of Forests shattered the optimism of April 2001. "I think the public was duped by this agreement," said Peter McAllister, "in large part by our colleagues who were desperate for 'a win'. We overstated the gains in this campaign." Turning Point co-chair Art Sterritt agreed that logging had undergone no significant change, and expressed First Nations' desire for "environmentally responsible practices" in the remaining coastal old-growth stands.[144]

Despite this setback, the Central Coast LRMP continued working through 2003 to hammer out a consensus among 17 groups that included government, First Nations, communities, workers, forest and mining companies, tourism interests, and environmentalists. They submitted their land-use plan in April 2004, although a final version rested on further talks between government and First Nations. The plan called for a reduction in the central coast's AAC from four million to three million cubic metres, significant withdrawals for protection of wildlife habitat, and the adoption of a management system based on ecological principles. Forest companies and contractors have been informed that they can expect compensation for lost cutting rights, but whether this is a vision that can successfully restructure economic and environmental relationships along the central coast remains to be seen.[145]

Even before receiving the document the Liberal government took steps to involve First Nations more fully in the central-and-north-coast forest industry. A $95-million revenue-sharing fund and timber rights taken from major coastal licensees provided the basis for agreements that may play a part in future treaty talks. Forests Minister Mike de Jong signed the first in a series of agreements with the Gitga'at First Nation late in 2003. In addition to providing $1.57 million over five years, the deal afforded the Gitga'at the rights to 290,000 cubic metres of timber around Hartley Bay. In signing, the Gitga'at had acknowledged that their "aboriginal interests in relation to forestry development within their traditional territory have been accommodated", de Jong remarked. He also expressed hopes that it would pave the way for other members of the Turning Point Initiative to sign similar accords. That proved to be the case as the Haisla, Heiltsuk, Kitasoo and Wuikinuxv First Nations have recently inked agreements involving $10.78 million in revenue sharing over five years and access to over a million cubic metres of timber in their regions. Each acknowledged that the benefits constituted a "workable interim accommodation of the economic component of their aboriginal interests regarding forestry decisions". Elsewhere along the coast, unresolved aboriginal land claims have continued to disrupt logging, and these interim agreements will provide only a temporary respite from future conflict as treaty negotiations continue to drag on.[146]

Abandoned steam-donkey boiler on the southern shore of Pitt Island, 2005.
Gavin Hanke photograph, RBCM.

Afterword

Up-Coast Up to Now

Having completed the main body of this book in the summer of 2004, I thought it appropriate to consider briefly events of the ensuing year along the central and north coast before sending this work off to press. It has been a busy year indeed in the region, involving the apparent demise of the Skeena Cellulose mill at Prince Rupert, court decisions that strengthened First Nations land and resource rights, further corporate reshuffling, and militant action by the Haida to force consultation and accommodation of their interests in the wake of the sale of Weyerhaeuser's coastal interests to Brascan. What I initially foresaw as a relatively straightforward summary of recent events instead became a complicated effort to capture a clear image of an industrial structure undergoing massive upheaval.

What will emerge out of the rubble of the old order is anybody's guess. Some will regret its passing, many will not. Forestry has provided at best a flimsy source of stability for most coastal communities. Although Prince Rupert and Terrace enjoyed a period of relative postwar prosperity, even that has crumbled in recent years, and the collapse of Skeena Cellulose has left their economies reeling. A recent regional analysis paints a grim picture: "poverty, poor social and health conditions, low morale, disaffected young people, inadequate housing, and lack of access to education and skills upgrading ... combined with high unemployment levels and limited business opportunities." Ongoing over-harvesting of the most valuable and accessible stands, the inevitable conversion to a second-growth economy, raw-log exports, and mill automation provide little hope for employment revival. Nor has a solution been found for the region's historical inability to capture the social and economic benefits of its resources. Logs from the most productive remaining sites continue to flow elsewhere for processing, leaving non-aboriginal coastal communities and First Nations the difficult task of building a new economy upon damaged ecosystems.[1]

Perhaps nothing over the year separating the summers of 2004 and 2005 captures this unravelling with greater clarity than the failure to resurrect Skeena Cellulose, a postwar linchpin of north-coast forestry. Early in 2004 New Skeena Forest Products sought protection under the Companies

Creditor Arrangement Act after the communities of Prince Rupert, Port Edward, Terrace and Hazelton began seizing equipment against unpaid taxes. Prince Rupert, having maintained its budget in anticipation of receiving nearly $8 million owed from the two previous years, slashed spending and raised residential taxes in an effort to cope with its fiscal crisis.[2]

Meanwhile, New Skeena continued to seek out capital needed to commence production. Expressions of interest from Tembec and the Woodbridge investment firm secured extensions from BC Supreme Court Justice Donald Brenner, but both backed out. In August, MarlinPatterson, a New York-based equity fund, investigated a short-term investment premised on the prospect of a quick sale. That, too, fell through, and on September 17 New Skeena CEO Daniel Veniez requested that the court put the firm into receivership. Brenner approved the application three days later, apparently setting in motion the liquidation of assets including the pulp mill, sawmills at Terrace and Hazelton, a whole-log chipper in Hazelton, logging operations, and Tree Farm Licence (TFL) No. 1. A "deeply saddened" Veniez cited falling pulp prices, a strengthening Canadian dollar, high-cost timber and Skeena's "longstanding reputation as a cash-sucking white elephant" as factors, but would later direct blame at "self-destructive and self-important bright lights of local government" who had created a climate hostile to investment.[3]

Prince Rupert, facing debt rising to $12 million and owed a reported $22 million in property taxes, could only hope that court-appointed receiver Ernst and Young would find a buyer willing to keep the Skeena Cellulose enterprise intact and operate the mills. Chances of recovering the back taxes were slim regardless of the outcome as the city lacked standing as a secured creditor; of more long-term importance were over 400 desperately needed jobs should the pulp mill reopen. Unemployment stood at 11.5 per cent in the northwest in 2004, well over the provincial average of 8.1 per cent, some 1,200 jobs had been lost in the previous five years, and the region's population had fallen nearly 7 per cent in the same period.[4]

The New Skeena era had admittedly been a "horrible emotional rollercoaster" for Prince Rupert, declared the *Daily News*, but when a Skeena-Queen Charlotte Regional District meeting produced only a gloomy scenario of lost jobs, lost taxes and cuts to services, the editor criticized civic leaders for a lack of leadership. "All this nay-saying drags our property values down, drives businesses away, convinces families not to move here, and makes otherwise positive, good-natured people hang their heads," he observed. Prince Rupert Mayor Herb Pond defended the boosterism of his council colleagues, cited a "can do" attitude at city hall, and pledged to achieve "the revitalization of the economy of the northwest".[5]

No group had more to lose from permanent closure of Skeena Cellulose than Local 4 of the Pulp, Paper and Woodworkers Union of Canada (PPWC). In mid October, with the deadline for submission of offers to the receiver looming, the union retained consultant Maurice Thibeault to assist in finding a buyer willing to take over and operate the assets. North Coast MLA Bill Belsey explored options with anxious delegations, but on November 22 Larry Prentice of Ernst and Young asked the court's approval to hire an auctioneer to liquidate the New Skeena holdings. Despite pressure from secured creditors for a quick sale, Prentice explained, he had sought an operator for the pulp mill. These efforts had failed, which in turn made it difficult to find a buyer for the sawmills. Lack of a local market for the chips produced in the process of cutting decadent north coast timber made profitable sawmill operation problematic. Prentice did offer one thin ray of hope. A group of local investors had begun negotiating for purchase of the Terrace mill. Vague feelers had also been received from Asian parties, but their interest apparently did not extend beyond buying parts of the pulp mill for dismantling and shipment to China. Liquidation seemed likely, therefore, with only the Terrace mill a serious candidate for survival.[6]

Two last-minute offers enabled a desperate City of Prince Rupert to obtain a one-week extension from Brenner in late November. Prentice, unhappy about the extra time granted for this "last-ditch effort" to keep New Skeena's assets intact, recommended disapproval of both offers. Hazelton mayor Alice Maitland, contemplating the loss of the Carnaby sawmill and whole-log chipper, called for the receiver to weigh the needs of communities against those of creditors. A decision to liquidate might lead to both facilities being broken up and sold off, leaving Hazelton with little opportunity for an economic recovery. "Our communities feel like they've paid for that equipment over and over again," Maitland said. "It needs to stay here."[7]

Maitland and other civic leaders were also worried about the fate of TFL No. 1. Would the tenure be sold off to a firm planning to process the fibre outside the region? The receiver offered some hope on this point, recommending approval of a $4.8 million offer from the Lax Kw'alaams Band Council of Port Simpson for the licence and Terrace log-sorting yard. That outcome promised local processing, but recent changes to forest legislation and First Nations challenges had created a good deal of uncertainty about the value and security of TFL tenure in the region. The Gitanyow, Metlakatla and Lax Kw'alaams First Nations had challenged the transfer of Skeena's TFL from the province to NWBC Timber and Pulp, arguing that it would impact forestry revenues on areas subject to land-claims negotiations.[8]

The Metlakatla and Lax Kw'alaams First Nations had subsequently signed interim agreements with the province giving them access to specified timber volumes and a share in revenues, while the Gitanyow Hereditary Chiefs continued with their legal challenge. That brought a measure of success in the form of a December 2002 BC Supreme Court ruling that the Crown had failed to consult and accommodate Gitanyow interests. The province appealed that decision, and court-ordered negotiations on an interim Forest and Range Agreement had subsequently failed. By the end of November 2004 the Gitanyow Hereditary Chiefs were back in court, their confidence buoyed by recent BC Supreme Court and Supreme Court of Canada decisions supporting the Haida First Nation's challenge to a similar tenure transfer on the Queen Charlotte Islands. Further discussion of that subject follows, but it seems clear that doubts surrounding the security of TFL No. 1 in the northwest would have cooled enthusiasm among potential corporate investors. When the BC Supreme Court ruled again in late December that the Crown had failed in its duty to consult in the transfer of Skeena Cellulose to NWBC the Gitanyow issued a "buyer beware" warning. The existence of "fundamental legal defect" in New Skeena's tenures within their territory implied considerable financial risk for potential investors.[9]

Other uncertainties flowed from an accumulation of "silvicultural liabilities" on TFL No. 1. Would a purchaser be held financially responsible for reforestation commitments unfulfilled by New Skeena? There were also the implications of the Liberal government's 2003 Forestry Revitalization Act to consider. That legislation called for a 20-per-cent reduction in the allowable annual cut of major licence holders for redistribution to First Nations, community forests, woodlots and a market-driven auction system. A multi-million dollar compensation fund had been established, but Prentice doubted that the purchaser of TFL No. 1 would be eligible for a share because the backlog of silvicultural liabilities likely exceeded the value of compensation permitted under the Act. Thus, uncertainty about the financial value of TFL No. 1 further complicated the tenure's status.[10]

Meanwhile, the liquidation process recommended by the receiver gained Brenner's approval in early December. The pulp mill and Carnaby facilities would be put up for sale while negotiations continued on the Terrace facility. Brenner also endorsed the Lax Kw'alaams purchase of TFL No. 1, although many issues remained to be ironed out. PPWC Local 4 Head Frank DeBartolo expressed pleasure at the prospective TFL outcome, but criticized the receiver for failing to support the union's efforts to find a pulp mill operator. A couple of candidates were still in play, he argued, and an extension to ensure a fibre supply might yet save the operation. About a third of the mill's former employees had found work, but many only at part-time low-wage jobs. Although Prince Rupert mayor Herb Pond seemed to consider further extensions unjustified since no party would commit to covering New Skeena's costs over the coming months, DeBartolo questioned the rush to satisfy the firm's creditors. "Do we really care if the secured creditors get a few dollars more?" he asked.[11]

Prince Rupert's fiscal problems deepened in early 2005. Another round of cuts to the city's budget came in January, affecting a wide range of services. But as citizens vented their anger and fears at city council meetings, efforts continued to cobble together a way to preserve New Skeena's industrial structure. In late February the PPWC and community of Lax Kw'alaams, which had established a company called Coast Tsimshian

Resources (CTR) to buy the TFL, announced the development of a business plan to preserve New Skeena's assets. But that deal would require consultation with other First Nations, along with Ministry of Forests approval. Also key to this initiative was the arrival of a Chinese delegation to study the pulp mill and fibre supply. China's rapidly expanding economy had triggered a global search for raw materials by government agencies such as the China Paper Group (CPG), whose appearance provided the first real glimmer of hope that the pulp mill might be saved. Prospects seemed brighter for preservation of the Terrace sawmill, as well. A group of local entrepreneurs had stepped forward with a serious offer. The first auction of New Skeena holdings was set to commence at Hazelton, though, and a reprieve from the courts was needed to slow the liquidation process.[12]

As February wound down, a series of meetings involving CPG, CTR representatives, Prince Rupert officials, the PPWC consulting group and the Terrace investors took place. Some two hundred residents gathered at the Terrace sawmill on the 23rd to celebrate the tentative purchase of that facility by the local group headed by John Ryan, contingent only upon acquisition of the land the plant occupied. Cautious optimism seemed the order of the day; if CTR succeeded in acquiring New Skeena's tenures, arrangements might be developed to supply the Terrace mill with logs, and if CPG bought and operated the pulp mill Terrace would have a local outlet for its chips.[13]

But would the Chinese operate the pulp mill, or dismantle it for shipment to China? Maurice Thibeault emerged from one meeting predicting the latter scenario. North Coast MLA Bill Belsey and Frank DeBartolo quickly refuted Thibeault's statements, Belsey pointing to a large deposit the Chinese had placed on the Watson Island assets. Their primary concern was a long-term fibre supply, he maintained, and CTR CEO Wayne Drury seemed optimistic that a "cooperative, integrated approach" by all interests could provide the fibre plan needed to rebuild the northwest forest industry.[14]

There the issue stood when the CPG delegation returned to China, promising to return later in March. When rumours circulated that CPG would apply for permits to bring in Chinese workers to dismantle the mill, Skeena-Bulckley Valley MP Nathan Cohen promised to work with immigration officials to have the request denied. Any sale to CPG should be contingent upon operation of the Prince Rupert plant, he went on to say. Prospects for a full-scale re-opening dimmed, however, when Belsey announced that the Chinese intended to dismantle the newer "B" mill's production line for operation in China. Plans to run the older "A" mill at the site were intact, giving hopes for some employment and tax revenue for Prince Rupert. Half a mill was better than none, after all, and postponement of the equipment auction scheduled for late April provided further grounds for hope.[15]

Prince Rupert received good news of a more tangible sort in mid April with the announcement of a $100 million plan to establish a world-class container port. "It seems as if, after a long run of businesses closing, we are in for a reversal of that trend," enthused the *Daily News*. Premier Gordon Campbell emphasized the economic benefits of almost 3,000 new jobs in port development during an election campaign swing later that month, an event attended by CPG representatives. Prince Rupert was "a gateway to the Pacific, a gateway to opportunity, a gateway to hope for the North," he declared, but said nothing definite about a CPG investment.[16]

If Prince Rupert citizens were disappointed at the absence of firm news about their pulp mill, many had no doubt grown weary of a process that seemed to have no end. Idle since 1998, the pulp mill had been a source of much false hope during the New Skeena era. Still, columnist Leanne Ritchie defended the extensive coverage by the *Daily News*, stating that the New Skeena assets retained great importance to the region's economic destiny. Terrace had fashioned a potential success story, recently confirmed with court approval of the sawmill land acquisition by the Ryan Group. CTR's effort to acquire the TFL showed promise. "So the story now is to watch the pieces changing hands," Ritchie concluded. "We are just starting to see how and if they will fit themselves back together and who the new players will be." Moreover, some of the participants in the game were northwestern interests, with roots in communities that owed "it to themselves to keep the forestry infrastructure in place and functioning".[17]

CPG's intentions remained a matter of intense speculation. Ritchie reported in late April that the company's original plan involved taking the mill apart for transport home, but the various regional interests continued to negotiate with the Chinese to retain at least some of the facility. CPG's purchase of 80 per cent of the Watson Island mill at this time raised the stakes. Submission of an offer for the land occupied by the mill raised hopes further, as did the signing of memorandums of understanding with CTR and PPWC Local 4. "Coming on the heels of the container announcement, it seems Prince Rupert can do no wrong," declared the *Daily News*. "With a world-class container port handling huge shipments from Asia, firmer-than-ever dreams of a reborn mill and a booming tourism industry, the only thing left to hope for is a reinvigorated fishing industry."[18]

As this manuscript goes to the publisher in the late summer of 2005, it would be nice to report a positive conclusion to the Skeena Cellulose story. The receiver has recommended acceptance of CPG's $3.3 million offer for the mill, Watson Island land and other assets. The court approved both that transaction and CTR's offer of over $4 million for TFL No. 1. CPG has established a new unit called Sun Wave Forest Products, and reached a preliminary taxation agreement with the City of Prince Rupert. Numerous obstacles remain, however. The deal does not appear to be contingent upon mill operation, in whole or part, and CPG insists upon being released from New Skeena's entire $25 million property-tax liability. The province objects to this provision, arguing that it contravenes federal bankruptcy law. Fibre supply and labour agreements remain to be finalized, and the entire package will ultimately require Chinese government approval. "We are a long way from banking on this," says Prince Rupert mayor Herb Pond, but it remains the best, and likely the only opportunity for a revived pulp mill.[19]

With court decisions creating increasing uncertainty about the security of forest tenure, and pressure on the Ministry of Forests to consult with First Nations on resource decisions, the Liberal government pushed ahead with its policy of negotiating Forest and Range Agreements. Through this strategy the government hopes to provide "a stable operating environment" for industry, improve the investment climate and enhance the value of forest resources. In exchange for economic benefits individual First Nations agree that the province is providing "workable accommodation of their interests". Twenty-four of these interim measures agreements had been signed by September 2004, involving revenue-sharing and tenure opportunities. First Nations taking part along the central and north coast included the Lax Kw'alaams, Gitga'at, Metlakatla, Kitselas/Kitsumkalum, Kitasoo, Heiltsuk, Haisla and Kitkatla peoples.[20]

But many First Nations harboured doubts about the rewards. The timber volumes involved are small, and there are concerns that the agreements may compromise the exercise of aboriginal title rights. More than 2,000 aboriginal people demonstrated at the provincial legislature in May 2004, some accusing the government of forcing agreements on poverty-stricken bands while employing stall tactics in treaty negotiations. One speaker said it was time for fair negotiations over land and resources, "not 'take-it-or-leave-it' accommodation deals that force us to give up the ability to exercise our rights for token economic benefits." Forests Minister Mike de Jong denied that such agreements undermined aboriginal rights, criticizing those leaders who chose to "fight and hurl words of abuse" rather than negotiate.[21]

While the provincial government and First Nations around the province entered into interim agreements, the Haida awaited a Supreme Court of Canada decision on their challenge to the transfer of TFL No. 39 from MacMillan Bloedel to Weyerhaeuser. Weyerhaeuser's 1999 takeover of the province's forestry giant, and the transfer of cutting rights to thousands of acres of Graham Island forests had been accomplished without consultation, triggering another round of legal action. In 2002 the Haida welcomed a B.C. Court of Appeal ruling that both the Crown and Weyerhaeuser owed them a duty to good faith consultation to seek workable accommodations "with respect to the granting [and transfer] of tenures, other alienation of resources and management of the land in question."[22]

The duty to consult had now involved a more meaningful, and from industry's perspective, more threatening duty to accommodate First Nations cultural and economic interests. Weyerhaeuser and the Crown promptly appealed. Part of the Appeal Court decision enabled the Haida and Weyerhaeuser to make application for interim rulings while the Supreme Court of Canada deliberated. A potential complication arose from the provincial government's spring 2003 Forestry Revitalization legislation. That package eliminated a longstanding requirement that the Minister of Forests consent to the transfer of major licences such as TFLs. Here the province sought, asserts a West Coast Environmental Law analysis, to limit or avoid its duty to consult First Nations on tenure awards, transfers and management. A tenure undergoing transfer can be cancelled if the minister decided that the transaction would "unduly restrict competition" in timber, log or chip markets, but the new statute was mute on the subject of aboriginal rights and title. TFLs had become "more like private property" by eliminating the Crown's right of transfer approval, coupled with other changes that shifted the balance of managerial authority from Ministry of Forests staff to company foresters.[23]

The Haida and other First Nations protested the new Forest Act on the grounds that the amendments had been developed without consultation, but the province's attempt to evade its duty to consult hit another snag in late September 2004. With the Supreme Court of Canada's decision on the TFL No. 39 case expected soon, and rumours circulating of Weyerhaeuser's intention to sell its coastal interests, the Haida scored another victory in the BC Supreme Court. Justice Stephen Kelleher ruled that the company could not transfer its licence without notifying the Haida of the identity of the buyer and the terms of any proposed transfer. This, in effect, extended the Crown's duty to consult and accommodate to industry. Haida lawyer Louise Mandell called the ruling a clear victory. "The most important part," she asserted, "is the duty [to consult] doesn't go away because the province legislates away its control of the situation. And if the province doesn't have it, the court is prepared to put it on the company."[24]

All attention now turned to the Supreme Court of Canada, which rendered a mid-November decision one First Nations leader described as "a bit of a mixed bag." While upholding the Crown's duty to consult and accommodate where development might impact a legitimate land claim, flatly denying the province's argument that no such obligation existed until aboriginal title had been proven, the Court ruled that this requirement did not extend to industry. Chief Justice Beverly McLaughlin wrote that the duty to consult and accommodate was "grounded in the honour of the Crown", which had no right to "cavalierly run roughshod over aboriginal interests where claims affecting these interests are being seriously pursued in the process of treaty negotiation and proof." But the Crown's honour could not be delegated to third parties such as Weyerhaeuser.[25]

Weyerhaeuser and industry generally welcomed the decision which freed the private sector from the onus of direct negotiations with First Nations. The Haida also cheered the judgement, which University of Victoria law professor Hamar Foster called "a pretty powerful assertion of aboriginal title and rights even before they've been proven in court." Industry had been let off the hook, Louise Mandell commented, but all doubt about the province's duty to consult had been removed.[26]

Rumours of Weyerhaeuser's plan to sell its coastal assets, including the troublesome TFL No. 39 which included much of Graham Island, continued to circulate in the aftermath of the court's decision. But would the Haida have a say in any transaction involving their traditional lands? Weyerhaeuser had no duty to consult them, and the Ministry of Forests had cynically surrendered its right to approve or deny TFL transfers. Untroubled by these legal questions Weyerhaeuser ended its brief stay on the B.C. coast in February 2005, agreeing to sell its entire holdings to Brascan Corporation for $1.4 billion. A shareholder in Western Forest Products, which had emerged as a major player from the ashes of Doman Industries' 2004 bankruptcy, Brascan had little apparent interest in Weyerhaeuser's Crown tenures. Much more attractive were the firm's 258,000 hectares of private timberland, free from log export restrictions and available for real-estate development or sale.[27]

As concerns about possible mill closures on Vancouver Island mounted, Queen Charlotte Islands residents pondered issues raised by the transaction. Port Clements, which had supported the Haida court action, and where much of the logging workforce was employed by contractors to Weyerhaeuser on TFL No. 39, seemed undismayed at the American firm's departure. "I'm going to approve this from the perspective that anybody's got to be better than the worst corporate citizen I've ever seen," Mayor Dale Lore said. Guujaaw, president of the Council of the Haida Nation (CHN), expressed resentment that the province had once again failed in its duty to consult and accommodate despite the recent Supreme Court judgement. A visiting Brascan official maintained that the firm's interest lay primarily in its 4,500 hetares of private land on the Charlottes, and that consultation was strictly a provincial responsibility.[28]

Haida grievances were not limited to the Weyerhaeuser/Brascan deal at this time. The CHN had developed a Land Use Vision for Haida Gwaii in late 2004 while the government's Land Use Planning Forum entered its final discussions for a long-range planning document. That would be submitted to the province and Haida for final negotiations, but by late 2004 the planning process was bogged down as the representatives of various interest groups struggled to reach a consensus on protection for fisheries, wildlife and forest ecosystems. Ultimately the forum's proposals would fall short of the principles laid out in the Land Use Vision, a document emphasizing the centrality of spirituality, respect and co-existence with nature to Haida culture.[29]

The Haida also expressed bitterness over Ministry of Forests approval of Husby Forest Products logging plans in a number of areas the Land Use Vision considered worthy of preservation for their cultural and ecological significance. Finally, the CHN accused Weyerhaeuser, which did not finalize its transaction with Brascan until May 31, 2005, of reneging on a 2002

Six-Point Agreement to reduce the rate of logging, preserve cedar for Haida cultural uses and introduce no more mechanical harvesters. Enjoying new freedoms under a relaxed forest practices code and less oversight from Ministry of Forests staff under the revamped Forest Act, companies were "behaving like it's their last days", Guujaaw said.[30]

Nor were the Haida alone in rejecting the historical pattern of forest-exploitation policy makers and outside corporations had fashioned on the Queen Charlotte Islands. Sorting out the range of public opinion on the Islands Spirit Rising protest that occurred in the spring of 2005 is beyond the scope of this account, but it seems clear that many non-aboriginal islanders agreed with its critique of the absence of local control over resources and economic development, and the almost century-long drain of timber for processing at southern mills. These shared sentiments had produced a March 2004 Protocol Agreement between the CHN and the municipalities of Port Clements and Masset to collaborate "in common cause" for the benefit of the islands. For too long decision making had been controlled by "off island regimes" with little or no interest in the well-being of their communities, the parties agreed. The harmonization of Haida and Crown titles might even provide an opportunity for cooperation and rational planning needed to create environmental, cultural and social sustainability. Discordant voices would be heard on the islands in early 2005 as loggers lost wages to the protest. But uncertainty and resentment blended with a cautious sense of potential empowerment for communities fed up with the sight of log-laden barges departing for Vancouver.[31]

Fittingly, the Islands Spirit Rising protest by the Haida and their supporters began on March 21, 2005, when a flotilla of fishing boats blockaded Weyerhaeuser barges in Masset Inlet. The following day, protesters established checkpoints on two logging roads near Port Clements and at Queen Charlotte City, preventing loggers from going to work. They erected signs bearing the slogan "Enough is Enough", vowing to remain until the government consulted with the Haida about the Weyerhaeuser sale.[32]

Haida representatives had explained their intention to disrupt Weyerhaeuser's wood supply to a tense meeting at the Port Clements community hall on the evening before the protest began. Many workers conveyed their sympathy with Haida concerns over the state of forest management and the lack of consultation on the Queen Charlottes, but questioned their sacrifice of wages for the protest. Some expressed a willingness to assist in the blockade of Weyerhaeuser barges, a tactic that would accomplish the main objective while allowing logging to proceed. The situation demanded strong action, Guujaaw countered, and taking power from the corporations that controlled TFL No. 39 could bring long-term benefits to all islanders. "We want to kick Weyerhaeuser in the butt on their way out the door and slam it before the other guy gets in," he declared. Port Clements mayor Dale Lore asserted his own support for the initiative, calling for cooperation from Port residents. Not only was this the "moral and right thing to do", Lore said, it was also in the community's self-interest given the influence the Haida would have on the future of islands logging. By the end of the meeting, reported the *Observer*, "many residents were giving an uneasy support to the road blockade."[33]

By the end of March about a hundred local loggers, now represented by the United Steelworkers after a September 2004 merger with the IWA, had missed a week's work as the protest continued to gain momentum. Local 2171 asked the CHN to remove the roadblocks and permit members to return to work. After meeting with affected workers, Local 2171 President Darrel Wong reported a consensus on the legitimacy of the Haida claim to title, but a wide range of opinion on the more immediate issue. "There were people that were furious, there were people on-side and supporting the Haida, and there were people that were afraid," Wong revealed. Entrepreneurs were also feeling the pinch, although the *Observer* recommended staying the course until concessions had been forced from the province "because business as usual is just not working anymore".[34]

Outside support for the Islands Spirit Rising protest gathered steam in mid April when a coalition of more than 30 labour unions, environmental organizations and First Nations groups sent an open letter to Premier Gordon Campbell condemning his government's policy toward the Haida. The B.C. Government Employees Union, PPWC, Hospital Employees Union, David Suzuki Foundation, Sierra Club, and Union of B.C. Indian Chiefs were among the interests attributing the conflict to the government's evasion of its court-mandated duty of honourable consultation.[35]

Negotiations to resolve the dispute began amid increasing national and international attention, pushing the government to the bargaining table. Dale Lore accompanied Haida leaders to Vancouver to meet with provincial officials, but his support for the Haida disturbed some Port Clements residents. A delegation of "concerned citizens" presented a petition bearing 60 to 65 signatures to the village council demanding that Lore stop representing the community on forestry issues, including the current protest and any Haida-provincial government negotiations. Lore, still convinced that Port Clements should support the Haida action out of justice and self-interest, attended only one meeting with Weyerhaeuser representatives, having been excluded from the main sessions at the insistence of provincial negotiators.[36]

Four days of provincial and Haida discussions produced a draft agreement on April 22. Guujaaw discussed the agreement at a Queen Charlotte City "Unity Feast" that evening, billed as an event to "bring islands people together". Funds raised through ticket donations were to be distributed to needy families. This would be the first of a series of community gatherings the Haida sponsored to explain the terms and implications of the memorandum of understanding signed on May 11. The Haida gained a wide range of concessions from the province. Fourteen areas identified in the Land Use Vision received protection for cultural or ecological purposes

until the end of 2008, or completion of the Land Use Plan. In all a reported 20 per cent of the islands would be preserved from logging, although
the parties agreed on a 30-day period to finalize protected areas. A decision on the sustainable harvest level for the Queen Charlotte Islands would
not come for another six months. There was also a commitment to develop a new approach to land-use planning based on ecosystem management
principles, to "connect land and resources to community viability, with the
intent to design a sustainable Island economy".[37]

The Islands Spirit Rising protest also produced immediate and long-
term financial benefits. A tenure with an annual volume of up to 120,000
cubic metres would be awarded to the Haida, along with an initial payment
of a $5-million share in resource revenue. The accommodations set out in
the agreement included no obligations or liabilities arising from previous
infringement of title and rights that courts might decide were owed the
Haida in the future, and imposed no limit on future treaty agreements. But
the province would have met any "accommodation obligations" arising
from the Supreme Court of Canada's recent Haida decision and the transfer of TFL No. 39. The door remained open for a Haida lawsuit against
Weyerhaeuser for alleged breach of its 2002 Six-Point Agreement.[38]

Although the Campbell government delayed signing the agreement
until the day after its re-election in spring 2005, finally bringing the protest
to an end, Guujaaw cautioned that many details remained to be ironed out.
Nor would the conclusion of those negotiations bring anything more than
another step in the land-use planning process that had been stalled by the
protest. Now, at least, provincial and Haida negotiations could begin discussions on the recommendations submitted by the planning forum. But
the finish would not come until "we know what the fate of the land will be,
that the timber economy will be managed responsibly, that the people are
able to live a good life here," Guujaaw claimed.[39]

As news of the agreement spread, some wondered if their access to the
good life would be curtailed. A Port Clements logging contractor worried
about the potential loss of cutting rights to the new Haida tenure. The government would protect the major licencees, but small-business loggers
faced an uncertain future. Whatever the outcome, it did not include
Weyerhaeuser, which officially departed the Queen Charlotte Islands on
May 29. Its successor, Brascan, split its new coastal holdings into two
entities. Island Timbers would operate the profitable private forestlands
concentrated on Vancouver Island, while Cascadia Forest Products took
responsibility for the ailing public lands. No longer would profits accumulated on private lands hide the poor performance of Crown forest assets,
Cascadia CEO Hugh Sutcliffe said, hinting at a major restructuring involving closure of some mills.[40]

That uncertainty, at least, did not extend to the Queen Charlottes,
where local wood processing had never figured in corporate and government policy. But the new Haida tenure and protection for more forestland

meant change. Some welcomed this as an opportunity to create a sustainable forest economy under local control with greater employment benefits for islands communities. Others were uneasy, especially the non-aboriginal entrepreneurs at Port Clements who had attempted to subvert Lore's support of the recent protest. One Port Clements resident predicted that the new agreement would "have a catastrophic effect on the economic well-being of the islands". Reports of Haida efforts to purchase TFL No. 39 raised the stakes even higher. "Everybody is worried about their own piece of the pie and rightly so," wrote a Tlell resident. "But everybody has to get over the fact that the mismanaged pie may one day belong to the Haida."[41]

A range of opinion also existed among the Haida, Guujaaw informed the audience at one of the post-protest public forums. Much remained to be decided, with a revived land-use planning process providing one pathway to arriving at the proper balance between ecology, culture and economy. The Islands Spirit Rising protest had been a necessary step in shaping the future, Guujaaw asserted. "People did a great service to Haida Gwaii. History will show it was the right thing to do."[42]

It will also be up to history, and future historians, to assess other aspects of recent changes along the central and north coasts. In analysing the status of northwest-coast forestry for Ecotrust's 2001 publication *North of Caution*, Ben Parfitt wrote that the region's communities shared a sense of abandonment. The mill closures and capital-flight of ensuing years have only confirmed the validity of that observation. But Parfitt also pointed to the forging of "new plans and new alliances" by the region's residents, generating hopes for sustainable forests and communities.[43]

The government's community forest licence initiative provides one new approach to achieving local control and benefits. Several up-coast communities have responded to the invitation to apply for tenures. Since receiving their invitation in October 2004 the Bella Coola Resource Society has devoted itself to identifying timber to support local business and value-added manufacturing. But the task has been complicated by an overabundance of low-value hemlock left by decades of logging, and the needs of other interests, including the Nuxalk First Nation, Interfor and independent loggers must be met. The society has identified several potential areas, but has yet to submit a formal application.[44]

Masset was also invited to apply, exciting local operators but raising Haida concerns over yet another instance of the government's failure to consult. Other Protocol Agreement communities wondered why they had been left out. Some saw the Masset offer as an attempt to divide Queen Charlotte Islands communities. In the end the CHN endorsed Masset proceeding with a tenure application. The volume in question, only 25,000 cubic metres a year, would create at least a few local jobs and was not worth the potential bickering obstruction might cause among the Protocol group. Prince Rupert and the District of Port Edward also initiated efforts to obtain a community forest licence in early 2005, but locating a suitable volume of timber will be problematic. Indeed, it seems unlikely that the timber volumes available under the community forest policy will generate considerable employment, and will certainly mean no restructuring of the up-coast forest industry. Still, the opportunity for at least a measure of local control strikes a strong chord in regions around the province.[45]

On the mid coast, the main hope for a new forest economy lies in the Great Bear Rainforest Agreement. Some First Nations' concerns have yet to be resolved, and impatience among environmental groups is growing at the slow pace of government ratification. Meanwhile, environmentalists charge that the consensus agreement forged with the five Coast Forest Conservation Initiative (CFCI) companies on protected areas and ecosystem-based management has masked a commitment to business-as-usual logging. An April 2005 annual report on forest practices by ForestEthics, Greenpeace, the Rainforest Action Network and Sierra Club of B.C. awarded the government and mid-coast operators failing grades for protection and ecological management. Government inaction and the lack of evidence of a shift to more ecologically sensitive forest practices threatened a "fragile five-year peace", Greenpeace's Catherine Stewart announced, warning that consumers of BC forest products would be notified of the lack of progress. The CFCI partners countered with a more positive April statement, looking forward to final government ratification of the consensus agreement by the autumn of 2005.[46]

In the final analysis, Ken Coates argues, the best hope for the northern economy rests upon First Nations treaty settlements. Treaties would provide millions of federal dollars, funds much more likely to remain for regional, economic and social development than multinational investment capital. In addition, elimination of the uncertainty surrounding land and resources would provide assurance to potential investors. The Nisga'a Treaty represents a promising development from this perspective, the potential first step in the creation of "a new kind of shared vision" of northern development. At the same time, as Coates acknowledges and this study demonstrates, resistance to the assertion of aboriginal rights remains strong in many quarters.[47]

The northwest coast's relatively brief and unequally distributed experience of postwar resource-based prosperity has come to an end. Even at its height, the region's position in the global economy provided an uncertain footing for its resource communities. The stories of Swanson Bay, Ocean Falls and, more recently, Prince Rupert reveal much about the vulnerabilities associated with B.C.'s faith in large-scale resource extraction under business-government cooperation, technocratic management and technological sophistication. It is abundantly clear that those who controlled the decisions from distant metropolitan centres lacked motivation to retain the wealth generated for regional well-being, and that the structures of technological and managerial control served only to pursue profitability in world markets at the expense of social and ecological sustainability.

What follows for coastal communities will be worked out in coming decades. One can only wish the best to those who face the difficult task of forging the new model, in the hope that they can create one more equitable, sound and balanced than the old structure. Having researched and written this history in Victoria, one based entirely on dry documentary evidence, I have gained some limited sense of the obstacles to be overcome. Making history is inevitably a much less tidy business than writing it, but as Ian Gill writes in the forward to *North of Caution,* a successful future for the coastal forest depends not just on the preservation of trees, but also on "options for people to live lives cf dignity and prosperity".[48]

Endnotes

Introduction

1. Paul Hirt, *A Conspiracy of Optimism: Management of the National Forests Since World War Two* (Lincoln: University of Nebraska Press, 1994).

Chapter 1: Shallow Roots

1. Robin Fisher, *Contact and Conflict: Indian-European Relations in British Columbia, 1774-1890* (Vancouver: UBC Press, 1977), 204-8; Paul Tennant, *Aboriginal Peoples and Politics: The Indian Land Question in British Columbia, 1849-1989* (Vancouver: UBC Press, 1990), 50-51.
2. Fisher, *Contact and Conflict*, 188, 205-11; E. Palmer Patterson II, "A Decade of Change: Origins of the Nishga and Tsimshian Land Protests in the 1880s," *Journal of Canadian Studies* 18 (Fall 1988), 40-54.
3. Jean Barman, *The West Beyond the West: A History of British Columbia* (Toronto: University of Toronto Press, 1991), 180.
4. G.W. Taylor, *Timber: History of the Forest Industry in B.C.* (Vancouver: J.J. Douglas, 1975), 7-8; Richard Mackie, *Trading Beyond the Mountains: The British Fur Trade on the Pacific, 1793-1843* (Vancouver: UBC Press, 1997), 203-17.
5. Robert A.J. McDonald, *Making Vancouver: Class, Status, and Social Boundaries, 1863-1913* (Vancouver: UBC Press, 1996), 7; Donald MacKay, *Empire of Wood: The MacMillan Bloedel Story* (Vancouver: Douglas & McIntyre, 1982), 14; Joseph C. Lawrence, "Markets and Capital: A History of the Lumber Industry of British Columbia, 1778-1952," (M.A. thesis, University of British Columbia, 1957), 24-27.
6. Gordon Hak, *Turning Trees into Dollars: The British Columbia Coastal Lumber Industry, 1858-1913* (Toronto: University of Toronto Press, 2000), 23, 69-72; Robert E. Cail, *Land, Man and the Law: The Disposal of Crown Lands in British Columbia, 1871-1913* (Vancouver: UBC Press, 1974), 100.
7. D.O.L. Schon, "Unique British Columbia Pioneer," *Forest History* 14 (Jan. 1971), 18-22.
8. Walter G. Hardwicke, *Geography of the Forest Industry of Coastal British Columbia* (Vancouver: Tantalus Research, 1963), 11-12; Ken Drushka, *Working in the Woods: A History of Logging on the West Coast* (Madeira Park: Harbour Publishing, 1992), 50-51.
9. Richard A. Rajala, *Clearcutting the Pacific Rain Forest: Production, Science, and Regulation* (Vancouver: UBC Press, 1998); "The Modern Flying Machine at Work," *Western Lumberman* 11 (July 1914), 40 [hereafter cited as *WL*].
10. Cole Harris and Robert Galois, "A Population Geography of British Columbia in 1881," in Cole Harris, *The Resettlement of British Columbia* (Vancouver: UBC Press, 1997), 147-50; Rolf Knight, *Indians at Work: An Informal History of Native Labour in British Columbia, 1858-1930* (Vancouver: New Star Books, 1996), 179-80.
11. James Andrew McDonald, "The Marginalization of the Tsimshian Cultural Ecology: The Seasonal Cycle," in Bruce Alden Cox, ed., *Native People, Native Lands: Canadian Indians, Inuit and Metis* (Ottawa: Carleton University Press, 1992), 202-3; Ken Campbell, "Hartley Bay, B.C.: A History," in Margaret Seguin, ed., *The Tsimshian: Images of the Past: Views for the Future* (Vancouver: UBC Press, 1984), 15; Janice Beck, *Three Towns: A History of Kitimat* (Kitimat Centennial Museum Association, 1983), 14-15.

12. John Lutz, "After the Fur Trade: The Aboriginal Labouring Class of British Columbia, 1849-1890," *Journal of the Canadian Historical Association* 3 (1992), 91.

13. Canada, *Annual Report of the Department of Indian Affairs for the Year Ended June 30, 1902* (Ottawa, 1903), 257; Knight, Indians at Work, 153, 234; Campbell, "Hartley Bay," 12.

14. Knight, *Indians at Work*, 232, 237; Ranger District 7, Annual Management Report, 1923, B.C. Department of Lands Records, Reel B9899, File 027637, B.C. Archives [hereafter cited as GR1441, BCA]; Canada, *Annual Report of the Department of Indian Affairs for the Year Ended June 30, 1893* (Ottawa, 1894), 21; Canada, *Annual Report of the Department of Indian Affairs for the Year Ended June 30, 1897* (Ottawa, 1898), 69; Canada, *Annual Report of the Department of Indian Affairs for the Year Ended June 30, 1902* (Ottawa, 1903), 257; Cliff Kopas, *Bella Coola: A Story of Effort and Achievement* (Vancouver: Mitchell Press, 1970), 263.

15. Hak, *Turning Trees*, 77-79.

16. Elizabeth Fay Kelley, "Aspects of Forest Resource Use Policies and Administration in British Columbia," (M.A. thesis, University of British Columbia, 1976), 69-70; R. Peter Gillis and Thomas R. Roach, *Lost Initiatives: Canada's Forest Industries, Forest Policy and Forest Conservation* (New York: Greenwood Press, 1986), 137-39.

17. H.N. Whitford and R.D. Craig, *Forests of British Columbia* (Ottawa: Commission of Conservation, 1918), 89; Hak, *Turning Trees*, 98-99, 112; Taylor, *Timber*, 78-79; "Selling Pulp Wood From the Crown Forests," nd., GR1441, Reel B4365, File 04009, BCA.

18. Whitford and Craig, *Forests*, 89-90; Hak, *Turning Trees*, 105.

19. Canada, *Annual Report of the Department of Indian Affairs for the Year Ended June 30, 1904* (Ottawa, 1905), 265-68; Canada, *Annual Report of the Department of Indian Affairs for the Year Ended March 31, 1907* (Ottawa, 1908), 242.

20. Richard A. Rajala, "There Was No Place Like It: The Rise and Fall of Ocean Falls, A British Columbia Paper Town, 1900-1985," (Ms prepared for the Canadian Museum of Civilization, December 2000), 4-5; *WL* 5 (June 1908), 19; *WL* 7 (Feb. 1909), 17.

21. Bruce Ramsay, *Rain People: The Story of Ocean Falls* (Ocean Falls Centennial Committee, 1971), 47-49; Rajala, "There Was No Place", 6; *WL* 7 (July 1910), 46; *WL* 8 (Apr. 1911), 30; "B.C. Pulp Company Has Six Millions Capital," *WL* 8 (July 1911), 69.

22. *Pacific Lumber Trade Journal* 9 (Dec. 1903), 18 [hereafter cited as *PLTJ*]; *PLTJ* 10 (June 1904), 22; *PLTJ* 10 (Jan. 1905), 20; *Pulp and Paper Magazine of Canada* 3 (Nov. 1905), 39 [hereafter cited as *PPMC*]; *PPMC* 4 (July 1906), 176-77; Christine Wozney, "Swanson Bay Pulp Mill", *Eurocan News* 7 (Summer 1980), 10.

23. *PPMC* 4 (Aug. 1906), 189; "Pulp in September," *WL* 5 (Feb. 1908), 16; "Extensive Pulp Enterprises," *WL* 5 (Aug. 1908), 18; "The Swanson Bay Pulp Limits," *WL* 5 (Dec. 1908), 19.

24. "Big Pulp Works at Swanson Bay," *WL* 6 (June 1909), 14; *PLTJ* 15 (Sept. 1909), 27; *PLTJ* 15 (Oct 1909), 24; *WL* 6 (Dec. 1909), 18-19; Ranger District Seven, Annual Management Report, 1923, GR1441, Reel B9899, File 027637, BCA.

25. *PPMC* 8 (July 1910), 163; "Swanson Bay Industry in New Hands," *WL* 7 (Aug. 1910), 19; "Extensive Additions to Swanson Bay Plant," *WL* 7 (Dec. 1910), 20; Ranger District Seven, Annual Management Report, 1923, GR1441, Reel B9899, File 027637, BCA.

26. *WL* 8 (Oct. 1911), 31; *PPMC* 9 (Dec. 1911), 420; *PPMC* 10 (Feb. 1912), 41; *PLTJ* 18 (Feb. 1913), 52; *WL* 12 (Oct. 1915), 20.

27. C.L. Barker, "History and Description of Pacific Mills Limited Mills at Ocean Falls, B.C.," (Unpublished manuscript, 1931, Royal British Columbia Museum), 4; "Ocean Falls Company Under New Management," *WL* 10 (Jan. 1913), 37; "Ocean Falls Closes Down," *PPMC* 11 (15 May 1913), 341; Ranger District Seven, Annual Management Report, 1923, GR1441, Reel B9899, File 027637, BCA.

28. J.A. Lower, "The Construction of the Grand Trunk Pacific Railway in British Columbia," in Thomas Thorner ed., *SA TS'E: Historical Perspectives on Northern British Columbia* (Prince George: College of New Caledonia Press, 1989), 344-50; George H. Buck, *From Summit to Sea: An Illustrated History of Railroads in British Columbia and Alberta* (Calgary: Fifth House, 1997), 66-71.

29. *PLTJ* 13 (Dec. 1907), 44; *WL* 5 (June 1908), 19; *PLTJ* 14 (July 1908), 22; *Timberman* 9

(Aug. 1908), 49; "Large Mill Closes Down," *WL* 5 (Oct. 1908), 18; "Fires Up North," *WL* 5 (Oct. 1908), 18; "Logs Tied Up," *WL* 5 (Dec. 1908), 20; *WL* 7 (Jan. 1909), 16.

30. *PLTJ* 14 (Nov. 1908), 30; *WL* 6 (Nov. 1908), 16; "Nass River Timber Limits Purchased," *WL* 7 (Sept. 1910), 24; *WL* 7 (June 1910), 24; *WL* 7 (Nov. 1910), 24; *WL* 9 (Jan. 1912), 42.

31. K.E. Luckhardt, "Prince Rupert: A Tale of Two Cities," in Thomas Thorner, ed., *SA TS'E: Historical Perspectives on Northern British Columbia* (Prince George: College of New Caledonia Press, 1989), 312; *WL* 8 (June 1911), 28; *WL* 8 (Aug. 1911), 33; "Will Operate Sawmill on Skeena River," *WL* 11 (May 1914), 33.

32. B.C., *New British Columbia: The Undeveloped Areas of the Great Central and Northern Interior* (Victoria: King's Printer, 1913), 64; N.J. Kerby, *One-Hundred Years of History – Terrace, B.C.* (Terrace Regional Museum Society, 1984), 15; WL 9 (July 1912), 33; "Lumber Trade of Northern B.C. Waters," WL 9 (Oct. 1912), 38; "A Pioneer Bulckley Valley Sawmill," WL 10 (Dec. 1913), 60; Buck, *From Summit to Sea*, 82.

33. "Coast Mill News," *WL* 5 (Mar. 1908), 14; "United Statsers Seeking Limits," *WL* 5 (June 1908), 19; *WL* 5 (July 1908), 28; "Another Big Sawmill," *WL* 5 (Sept. 1908), 16-17.

34. *WL* 7 (May 1910), 45; *WL* 7 (June 1910), 24; *WL* 7 (Nov. 1910), 24; *WL* 7 (Nov. 1910), 18; *WL* 8 (Oct. 1911), 31.

35. *WL* 6 (July 1909), 19; *WL* 6 (Nov. 1909), 16; *WL* 8 (Apr. 1911), 25; *PPMC* 9 (June 1911), 156; *WL* 8 (Oct. 1911), 31; *WL* 8 (Oct. 1911), 30; *WL* 8 (Dec. 1911), 30; *WL* 8 (Dec. 1911), 29; "Timber Wealth of Queen Charlotte Islands," *WL* 9 (June 1912), 37.

36. Campbell, "Hartley Bay," 18; *PLTJ* 14 (Nov. 1908), 30; "Hartley Bay Lumber Company Sells Out," *WL* 5 (Dec. 1908), 20; *WL* 6 (July 1909), 18; *WL* 9 (Mar. 1912), 43.

37. Kopas, *Bella Coola*, 265; *Bella Coola Courier*, 6 Dec. 1913, 1 [hereafter cited as *BCC*]; *BCC*, 10 Jan. 1914, 1; *BCC*, 24 Jan. 1914, 1; *BCC*, 7 Feb. 1914, 1.

38. H.S. Irwin to Chief Forester, 14 July 1913, GR1441, Reel B4351, File 0772, BCA; H.S. Irwin, List of Lumber and Shingle Manufacturers, Prince Rupert District, 21 Oct. 1913, Ibid.; "Namu," *BCC*, 5 Apr. 1913, 1; *BCC* 4 Oct. 1913, 1; *BCC*, 6 Dec. 1913, 1.

39. Thomas R. Roach, *Newsprint: Canadian Supply and American Demand* (Durham: Forest History Society, 1994), 2; Rajala, "There Was No Place", 8.

40. See Robert Howard Marris, "Pretty Sleek and Fat: The Genesis of Forest Policy in British Columbia, 1903-1914," (M.A. thesis, University of British Columbia, 1976), 43-67; Stephen Gray, "The Government's Timber Business: Forest Policy and Administration in British Columbia, 1912-1928," *BC Studies* 81 (Spring 1989), 26-28.

41. "Timber Policy of Government," *Victoria Daily Colonist* 24 Jan. 1912; "British Columbia Timber Lands," *American Forestry* 19 (July 1913), 447.

42. B.C., *Report of the Forest Branch of the Department of Lands for the Year Ending December 31, 1913* (Victoria: King's Printer, 1914), 1, 16.

Chapter 2: The Spruce Drive

1. B.C. *Report of the Forest Branch of the Department of Lands For the Year Ending December 31, 1914* (Victoria: King's Printer, 1915), 5-6, 24; Lower, "The Construction," 353-55; Luckhardt, "Prince Rupert," 314; "Prince Rupert Board of Trade," *WL* 12 (Jan. 1915), 21.

2. F.S. Wright, "The Lumber Industry of Central British Columbia," *WL* 12 (Apr. 1915), 15-17.

3. *WL*, 12 (Feb. 1915), 25; *WL*, 12 (Sept. 1915), 28; *WL* 12 (Nov. 1915), 22; *WL* 13 (July 1916), 35; Wright, "The Lumber Industry," 16; H.S. Irwin to Chief Forester, 29 April 1914, GR1441, Reel B4351, File 0772, BCA.

4. B.C., *Final Report of the Royal Commission on Timber and Forestry, 1909-1910* (Victoria: King's Printer, 1910), 55-56; *BCC*, 7 March 1914, 1; *BCC*, 25 April 1914, 2.

5. H.S. Irwin to Chief Forester, 12 May 1914; Acting Chief Forester to Irwin, 19 May 1914, GR1441, Reel B4351, File 0772, BCA; B.C., *Report of the Forest Branch of the Department of Lands for the Year Ending December 31, 1914* (Victoria: King's Printer, 1915), 15.

6. *BCC*, 24 Jan. 1914, 2.
7. See Ken Coates, "Divided Past, Common Future: The History of the Land Rights Struggle in British Columbia," in Roslyn Kunin, ed., *Prospering Together: The Economic Impact of Aboriginal Title Settlements in B.C.* (Vancouver: The Laurier Institution, 1998), 1-20; R.M. Galois, "The Indian Rights Association, Native Protest Activity and the "Land Question" in British Columbia, 1903-1916," *Native Studies Review* 8 (1992), 1-19; Knight, *Indians at Work*, 110.
8. Canada, Commission on Indian Affairs for the Province of British Columbia, Transcripts, Bella Coola Agency, GR1995, Reel B1454, BCA.
9. G. Harrison to Indian Agent, Masset, 30 Nov. 1914; Indian Agent to Secretary, Department of Indian Affairs, 4 Dec. 1914; Canada, Department of Indian Affairs, British Columbia Records, Reel B5642, BCA [hereafter cited as RG123].
10. Charles C. Perry to Secretary, Royal Commission on Indian Affairs, 20 Jan. 1916, GR123, Reel 5650, BCA.
11. "Ocean Falls Plants Will Be Started Up," *WL* 12 (Jan. 1915), 19; "Great Ocean Falls Plant Will Again Operate," *WL* 12 (Sept 1915), 32.
12. *BCC*, 17 July 1915, 1; *BCC*, 4 Sept. 1915, ; *BCC*, 11 Mar. 1916, 1; "Ocean Falls News," *BCC*, 21 Oct. 1916, 1; *BCC*, 1 July 1916, 1; Rajala, "There Was No Place," 8-9; Ramsay, *Rain People*, 103-4.
13. J.H. Hamilton, "The Pulp and Paper Industry in B.C.," *Industrial Progress and Commercial Record* 5 (July 1917), 328-30; *WL* 14 (July 1917), 28; "Suggest Unity of Purpose Between Workers," *B.C. Federationist*, 21 Dec. 1917, 3; "Great Pulp and Paper Plant of Pacific Mills," *WL* 15 (Mar. 1918), 40.
14. "Swanson Bay Pulp Mills Sold," *WL* 13 (Sept. 1916), 22; *WL* 13 (Oct. 1916), 23; *WL* 14 (Mar. 1917), 24; *WL* 14 (July 1917), 27; "Jas. Whalen Active in Pulp Circles," *PPMC* 27 (22 May 1919), 489.
15. "B.C. Pulp Company Increases Capitalization," *WL* 15 (Sept. 1918), 49; *WL* 15 (Dec. 1918), 29; "Jas. Whalen Active," 489; "A Good Showing for B.C. in Japan," *PPMC* 17 (12 June 1919), 555.
16. Donald MacKay, *Empire of Wood: The MacMillan Bloedel Story* (Vancouver: Douglas & McIntyre, 1982), 36-39; C.J. Taylor, *The Heritage of the British Columbia Forest Industry: A Guide for Planning, Selection and Interpretation of Sites* (Ottawa: Environment Canada, 1987) 110, 120-21.
17. "Spruce Mills Are Operating Up North," *WL* 13 (Jan. 1916), 27; "Northern Spruce Mills Get Shipping Priviledges," *WL* 13 (July 1916), 37.
18. "Activity in Northern Spruce," *WL* 14 (Mar. 1917), 30; "The Embargo on Aeroplane Spruce," *WL* 14 (Mar. 1917), 12.
19. *WL* 13 (Dec. 1916), 26; *WL* 14 (Jan. 1917), 24; "Northern Mills to Have First Monorail System," *WL* 14 (July 1917), 30; "Lumbering Activity in Northern B.C.," *WL* 14 (Nov. 1917), 31.
20. "Queen Charlotte Timber Being Marketed," *WL* 14 (May 1917), 28; *WL* 14 (July 1917), 28; "New Logging Company Extending Operations," *WL* 14 (Sept. 1917), 30.
21. "B.C. Spruce for Aeroplanes," *WL* 14 (Oct. 1917), 32; B.C., *Report of the Forest Branch of the Department of Lands for the Year Ending December 31, 1918* (Victoria: King's Printer, 1919), 5.
22. B.C., *Annual Report of the Forest Branch 1918*, 6; Ken Drushka, *HR: A Biography of H.R. MacMillan* (Madeira Park: Harbour Publishing, 1995), 106-7.
23. *WL* 14 (Nov. 1917), 30; *WL* 14 (Dec. 1917), 30; *WL* 15 (Mar. 1918), 34, 35; "Production of Aeroplane Spruce is Soaring," *WL* 15 (Aug. 1918), 39-40.
24. "Getting Busy on Spruce Contracts," *WL* 15 (Mar. 1918), 52; "Valuable Equipment for Spruce Timber Camps," *WL* 15 (June 1918), 39; "The Aeroplane Spruce Campaign," *WL* 15 (June 1918), 50; *WL* 15 (June 1918), 33; Drushka, *Working in the Woods*, 87-89; *WL* 15 (Feb. 1918), 42.
25. *WL* 15 (June 1918), 32; "Production of Aeroplane Spruce is Soaring," *WL* 15 (Aug. 1918), 39-40; *WL* 18 (Aug. 1918), 36.

26. *WL* 15 (May 1918), 34; *WL* 15 (June 1918), 25; *WL* 15 (Sept. 1918), 38; "The Y.M.C.A.
in British Columbia," *WL* 15 (Oct. 1918), 53; *WL* 15 (Oct. 1918), 58; "King Spruce and
His Realm," *Resources* 1 (Sept. 1918), 3,5; "Masset Inlet Timber Company's New Mill,"
WL 15 (Sept. 1918), 46; "Graham Island Spruce and Cedar Co. Ltd.," *WL* 15 (Oct. 1918),
79; "Spruce Mills on the Queen Charlotte Islands," *WL* 15 (Dec. 1918), 45; "B.C.'s Drive
for Airplane Spruce," *Timberman* 20 (Jan. 1919), 36-37.
27. "Lumbering Activity in Northern B.C.," *WL* 14 (Nov. 1917), 31; *WL* 15 (Feb. 1918), 42;
"Spruce Mills Along G.T.P. Railway," *WL* 15 (Apr. 1918), 38; *WL* 15 (May 1918), 33; *WL*
15 (Aug. 1918), 37; *WL* 15 (Oct. 1918), 57; "Sawmills Around Prince Rupert," *Resources* 1
(Sept. 1918), 5.
28. *WL* 15 (June 1918), 32; "Aeroplane Spruce Operations to be Curtailed," *WL* 15 (Jan.
1919), 35.
29. "Aeroplane Spruce Operations," 35-36.
30. *WL* 15 (May 1918), 37; B.C., *Report of the Forest Branch of the Department of Lands, 1918*, 6-7;
"Lumbering Activity," 31; "Production of Aeroplane Spruce," 40; *WL* 15 (Oct. 1918), 58.
31. *WL* 16 (Feb. 1919), 39; *WL* 16 (Feb. 1919), 39; "Spruce Department Now Balancing the
Ledger," *WL* 16 (June 1919), 48.

Chapter 3: The Twenties

1. *Pacific Coast Lumberman* 4 (Feb. 1920), 51 [hereafter cited as *PCL*]; "Leases Big Plant," *WL*
16 (May 1919), 32; "Prince Rupert Mill is a Hive of Activity," *WL* 16 (Aug. 1919), 30;
"Lumber Industry Booming," *Resources Monthly* 2 (Jan. 1920), 3; "Old Warhorse Gets Into
Game Again," *WL* 4 (Mar. 1920), 39-40; *PCL* 4 (Apr. 1920), 45; *PCL* 4 (June 1920), 76.
2. "Lumber Industry Booming," 3; "Big Development of Natural Resources," *Resources
Monthly* 2 (Jan. 1920), 3; *PCL* 4 (Apr. 1920), 45; *PCL* 4 (June 1920), 44; "Water-Wheel is
Used to Drive Sawmill Machinery," *WL* 16 (Sept. 1919), 31; *PCL* 4 (Aug. 1920), 63.
3. Buck, *From Summit to Sea*, 89; "Englemann and Sitka Spruce Make Northern B.C.
Famous," *WL* 17 (Oct. 1920), 66; Jack Mould, *Stump Farms and Broadaxes* (Saanichton:
Hancock House, 1976), 23-24.
4. Andrew Neufeld and Andrew Parnaby, *The IWA in Canada: the Life and Times of an Indus-
trial Union* (Vancouver: New Star Books, 2000), 25; Gordon Hak, "British Columbia
Loggers and the Lumber Workers Industrial Union, 1919-1972," *Labour Le Travail* 23
(Spring 1989), 72; *PCL* 4 (June 1920), 44; *PCL* 4 (July 1920), 45; E.C. Manning to Chief
Forester, 13 Jan. 1925, GR1441, Reel B9899, File 027637, BCA.
5. "Englemann and Sitka Spruce," 65; "Sitka Spruce is Now Permanent Factor," *WL* 17
(Aug. 1920), 61; *PCL* 4 (Oct. 1920), 59; "Masset Inlet Mill Closer," *PCL* 4 (Nov. 1920),
66; "High Freight Rates Stopped Shipment of Lumber," *WL* 17 (Dec. 1920), 42; Manning to Chief Forester, 13 Jan. 1921.
6. *WL* 16 (June 1919), 33; *WL* 16 (Aug. 1919), 27; *WL* 16 (Sept. 1919), 50; *WL* 17 (June
1920), 49; *PCL* 4 (Oct. 1920), 59; *PCL* 5 (Mar. 1921), 38; *PCL* 5 (May 1921), 76; *PCL* 5
(June 1921), 59; *WL* 18 (July 1921), 29; *PCL* 5 (Sept. 1921), 40; *WL* 19 (June 1922), 47.
7. Lukhart, "Prince Rupert," 312; "Englemann and Sitka Spruce," 65; *WL* 18 (Jan. 1921), 40;
Manning to Chief Forester, 13 Jan. 1921; *PCL* 5 (Jan. 1921), 42.
8. *PCL* 5 (Mar. 1921), 38; "Why is Western Hemlock Banned for Ties?", *WL* 18 (Mar. 1921),
37; *WL* 18 (July 1921), 29; *PCL* 5 (June 1921), 59; *PCL* 5 (Oct. 1921), 40; *PCL* 5 (Aug.
1921), 40.
9. E.C. Manning to Chief Forester, 9 Jan. 1922, GR1441, Reel B9899, File 027637, BCA;
"Cut Five Million Feet of Spruce Last Year," *WL* 19 (Mar. 1922), 71; For accounts of
LWIU decline see Hak, "British Columbia Loggers," 82-88; Richard A. Rajala, "Bill and
the Boss: Labour Protest, Technological Change and the Transformation of the West
Coast Logging Camp, 1890-1930," *Journal of Forest History* 33 (Oct. 1989), 168-79.

10. *PCL* 5 (Sept. 1921), 40; *WL* 18 (Oct. 1921), 29; *PCL* 5 (Oct. 1921), 82; *PCL* 5 (Nov. 1924), 40; Manning to Chief Forester, 9 Jan. 1922.

11. Manning to Chief Forester, 9 Jan. 1922

12. *WL* 19 (June 1922), 47; *WL* 19 (Nov. 1922), 29; *PCL* 7 (Mar. 1923), 49; "B.C. Timber for Los Angeles," *WL* 20 (May 1923), 36; George P. Melrose, "Scouting on the Queen Charlottes," *WL* 20 (July 1923), 19, 23; "Forests of Queen Charlotte Islands Yield World's Most Magnificent Aeroplane Spruce," *WL* 20 (Sept. 1923), 36-37, 40; *PCL* 7 (Dec. 1923), 36; *British Columbia Lumberman* 8 (Feb. 1924), 35 [hereafter cited as *BCL*].

13. *PCL* 6 (May 1922), 36; *PCL* 7 (Apr. 1923), 54; *PCL* 7 (Nov. 1923), 36.

14. J. Leslie, "The Skeena and its Timber," *PCL* 7 (Feb. 1923), 48; "Northern Lumbermen Hold Important Meeting," *PCL* 7 (Mar. 1923), 32.

15. Gray, "The Government's Timber Business," 47-48; *PCL* 7 (Apr. 1923), 54.

16. *PCL* 7 (Apr. 1923), 54.

17. "Royal Mill is a Steady Producer," *PCL* 8 (Jan. 1924), 62; "Terrace Active in Timber Industry," *PCL* 7 (Dec. 1923), 44.

18. "The Importance of the North," *PCL* 7 (May 1923), 22; *PCL* 7 (May 1923), 34; *PCL* 7 (July 1923), 61; *PCL* 7 (Aug. 1923), 34; *PCL* 7 (Nov. 1923), 36; *PCL* 7 (Dec. 1923), 36; "Prince Rupert Ships Lumber," *WL* 20 (Dec. 1923), 53.

19. "Northern B.C. Activities," *PCL* 7 (June 1923), 43; *PCL* 7 (Apr. 1923), 54; *BCL* 8 (Jan. 1924), 34.

20. "Northern B.C. Activities," 43; *BCL* 8 (Feb. 1924), 34; P.S. Bonney to Chief Forester, 7 June 1923, GR1441, Reel B4351, File 0772, BCA.

21. *BCL* 8 (Apr. 1924), 32; *BCL* 8 (May 1924), 32-33; *BCL* 8 (June 1924), 40; *BCL* 8 (Sept. 1924), 28; G.A. Hunter, "Pole Cutting a Growing Industry," *BCL* 8 (May 1924), 56; P.S. Bonney, Prince Rupert Forest District, Annual Management Report, 1924, GR1441, Reel B9899, File 027637, BCA [hereafter cited as PRFD, AMR].

22. R.L. Mavius, "Ollie Hanson is Successful Northern Pole Merchant," *WL* 21 (Apr. 1924), 20; G.A. Hunter, "Pole, Post and Tie Industry of Prince Rupert and District," *BCL* 8 (Oct. 1924), 39; Bonney, Annual Management Report, 1924.

23. G.A. Hunter, "The Northern Lumber Industry," *BCL* 8 (Oct. 1924), 57; *BCL* 8 (Aug. 1924), 28; *BCL* 8 (Sept. 1924), 28; *BCL* 8 (Oct. 1924), 37; *BCL* 8 (Nov. 1924), 30; *BCL* 8 (Dec. 1924), 30; *BCL* 9 (Mar. 1925), 32; *BCL* 9 (May 1925), 28; *BCL* 9 (Oct. 1925), 28.

24. "Hemlock Shipments from Prince Rupert," *BCL* 9 (Feb. 1925), 54; PFRD, AMR, 1924.

25. *BCL* 9 (Jan. 1925), 30; *BCL* 9 (Feb. 1925), 28; *WL* 18 (Mar. 1921), 31; *WL* 18 (May 1921), 35; "Pulp and Paper Items," *PCL* 5 (June 1921), 34; *PCL* 5 (June 1921), 59; "Prince Rupert Company Buys Timber," *WL* 18 (July 1921), 32; "Prince Rupert Company Makes Progress," *WL* 18 (July 1921), 29; *PPMC* 20 (10 Aug. 1922), 682; *WL* 19 (Oct. 1922), 44; Luckhardt, "Prince Rupert," 315.

26. *BCL* 8 (Dec. 1924), 29; *BCL* 9 (Jan. 1925), 30; "Sale of Prince Rupert Mill," *BCL* 9 (Jan. 1925), 22; *BCL* 9 (Mar. 1925), 32; "Mill to Specialize in High-Grade B.C. Spruce," *BCL* 9 (Apr. 1925), 57.

27. *BCL* 9 (Mar. 1925), 31, 32.

28. Chas. Piers, "Industry Conditions on North Coast of B.C.," *BCL* 9 (Sept. 1925), 42, 45; *BCL* 9 (Sept. 1925), 30; G.A. Hunter, "The Northern Sitka Spruce Industry," *BCL* 9 (Oct. 1925), 75-76.

29. "The Whalen Company's New Officials," *PPMC* 17 (26 June 1919), 504; "Financial Interests Behind Whalen Co.," *PPMC* 17 (10 Oct. 1919), 548; "Whalen Company Expands," *PPMC* 17 (17 July 1919), 592.

30. *PCL* 4 (Jan. 1920), 56; "Whalen Company's Car Ferry," *PPMC* 18 (12 Jan. 1920), 176; "Big Pulpwood Company Has Prosperous Future," *WL* 17 (Aug. 1920), 71-72; "Whalen Mills Had Good Year," *PCL* 4 (Aug. 1920), 95; Whalen Pulp and Paper Mills, Year 1920, GR1441, Reel B9899, File 027637, BCA.

31. "Whalen Doing Well," *PPMC* 19 (21 Apr. 1921), 431; *PCL* 5 (July 1921), 62; *WL* 18 (Sept. 1921), 20; *PCL* 5 (Sept. 1921), 40; *PCL* 5 (Nov. 1921), 40; E.C. Manning to Chief

Forester, 9 Jan. 1922, GR1441, Reel B9899, File 027637, BCA; *PPMC* 20 (16 Feb. 1922), 129; *PCL* 6 (Apr. 1922), 25; *PCL* 7 (Jan. 1923), 37; *PCL* 7 (May 1923), 34; *PCL* 7 (June 1923), 46; *PCL* 7 (Aug. 1923), 34.

32. "Pulp Company Ensures Water Supply," *WL* 20 (Sept. 1923), 45; "Whalen Mills Suffer With Japs," *PPMC* 21 (13 Sept. 1923), 918, 927.

33. Ranger District Seven, Annual Management Report, 1923, GR1441, Reel B9899, File 027637, BCA.

34. "Receiver Appointed for Whalen Pulp and Paper Co. Ltd.," *PCL* 7 (Oct. 1923), 31; *PCL* 7 (Nov. 1923), 51; *PCL* 7 (Dec. 1923), 36; *PPMC* 21 (27 Dec. 1923), 1268; *BCL* 8 (Feb. 1924), 35.

35. *BCL* 8 (Apr. 1924), 33; *BCL* 8 (May 1924), 33; *BCL* 8 (Oct. 1924), 37; "Whalen Conditions Improving," *PPMC* 22 (25 Sept. 1924), 1003; *BCL* 8 (Nov. 1924), 30; "Whalen Pulp Mill Winding Up," *PPMC* 23 (19 Mar. 1925), 304; *BCL* 9 (Mar. 1925), 32; "Whalen Assets to be Realized," *BCL* 9 (Apr. 1925), 29; *BCL* 9 (June 1925), 28.

36. "Whalen Pulp Assets Sale," *PPMC* 23 (14 June 1925), 630; "New Financial Arrangement for Whalen Co.," *Canada Lumberman* 45 (1 July 1925), 53; "Whalen Company Will Be Reorganized," *Canada Lumberman* 45 (1 Dec. 1925), 42; *PPMC* 24 (7 Jan. 1926), 19.

37. "Whalen Mills Now British Columbia Pulp and Paper Co.," *PPMC* 24 (28 Jan. 1926), 114, 123-24; "British Columbia Pulp and Paper Co. Makes More Pulp," *PPMC* 24 (30 Dec. 1926), 1603; *BCL* 11 (Apr. 1927), 21; *BCL* 12 (Mar. 1928), 32; G.A. Hunter, "The Prince Rupert District in 1928," *BCL* 13 (Mar. 1929), 26; Wozney, "Swanson Bay," 11; Bill Moore, "The Town That Vanished," *BCL* 58 (July 1974), 74-75.

38. A.J. Taylor, "Industrial Housing as Related to Isolated Manufacturing Plants," *Industrial Progress and Commercial Record* 6 (Aug. 1918), 79-80 [hereafter *IPCR*]; Gilbert A. Stelter and Alan F.J. Artibise, "Canadian Resource Town in Historical Perspective," in *Shaping the Urban Landscape: Aspects of the Canadian City Building Process* eds. Stelter and Artibise (Ottawa: Carleton University Press, 1982), 413-34; R. Robson, "Government Policy Impact on the Evolution of Forest-Dependent Communities in Canada Since 1880," *Unasylva* 47 (1996), 53-59; Roger A. Roberge, "Resource Towns: The Pulp and Paper Communities," *Canadian Geographical Journal* 94 (Feb./Mar. 1977), 28-35.

39. J. MacKellar, "History of Ocean Falls in Pictures" (unpublished manuscript, Volume 1, Folder 1, BCA); Roy Nakagawa, "Ocean Falls Recollection: A Story of the Town Where I Was Born," (unpublished manuscript, 1995 Royal British Columbia Museum), 44.

40. Howard B. Phillips, "The Days of My Years" (unpublished manuscript, BCA), 29-32.

41. J.H. Bradbury, "New Settlements Policy in British Columbia," *Urban History Review* 2 (Oct. 1979), 54; "A Typical 'Company' Town in B.C.," *B.C. Federationist*, 8 Sept. 1916, 3.

42. "Organized Movement for Abolition of Company Towns," *B.C. Federationist*, 13 Apr. 1917, p. 1; "Closed Towns Will Be Thing of Past," *Daily Colonist*, 29 June 1917, 7; *WL* 15 (May 1918), 23; Bradbury, "New Settlements Policy," 55-56; "Board of Trade Visits B.C. Pulp Mills," *PPMC* 17 (4 Sept. 1919), 757; J.S. Marshall, "Life at Ocean Falls," in "The Ocean Falls Story," Vol. 1, np; Rajala, "There Was No Place," 12.

43. *WL* 16 (Apr. 1919), 34; "British Columbia Notes," *PPMC* 17 (11 Dec. 1919), 1075; "Transatlantic Shipments," *PPMC* 18 (5 Feb. 1920), 142; *PCL* 5 (July 1921), 34; *PCL* 5 (Aug. 1921), 32; "Pulp and Paper as Factor in the Timber Industry of British Columbia," *WL* 19 (May 1922), 42; "Busy Times at Ocean Falls," *WL* 20 (Aug. 1923), 29; "Thirty Years of Papermaking: A Saga of Progress," *Paper People* 1 (July 1947), 5; Ramsay, *Rain People*, 93-101; Roach, *Newsprint*, 6.

44. MacKeller, "History of Ocean Falls in Pictures," Vol. 2, Folder 1, BCA; Richard A. Rajala, "Pulling Lumber: Indo-Canadians in the British Columbia Forest Industry, 1900-1998," *British Columbia Historical News* 36 (Winter 2002/2003), 2-13; Rajala, "There Was No Place," 13-15.

45. Drushka, *Working in the Woods,* 90; Marshall, "The Story of Ocean Falls," Vol. 1, np; "Industrial Railways in British Columbia," *PCL* 1 (Jan. 1920), 38.

46. "Selling Pulp Wood from the Crown Forests," 5-7.
47. Gray, "The Government's Timber Business," 43-44; *WL* 19 (June 1922), 47; "Northern B.C. Activities," *PCL* 7 (June 1923), 43; "Pacific Mills Ltds.," *BCL* 8 (Apr. 1924), 90; *BCL* 8 (May 1924), 33; G.A. Hunter, "The Prince Rupert District in 1928," *BCL* 13 (Mar. 1929), 26.
48. E.C. Manning to Chief Forester, 9 Jan. 1922, GR1441, File 9899, File 027637, BCA.
49. PRFD, AMR, 1922, GR1441, Reel B9899, File 027637, BCA; Ranger District Seven, Annual Management Report, 1923, GR1441, Reel B9899, File 027637, BCA; James Sirois, *Kimsquit Chronicles: Dean River, British Columbia* (Hagensborg: Skookum Press, 1996), 62-63; Drushka, *Working in the Woods*, 90; Gordon Robinson, "The Original Kitimat," *Northwest Digest* 14 (June 1958), 30; Beck, *Three Towns*, 26.
50. PRFD, AMR, 1923; Ranger District Seven, Annual Management Report, 1923; G.W. Nickerson, "Timber Resources of the Coast District of B.C. Tributary to Prince Rupert," *BCL* 8 (June 1924), 46.
51. Manning, Management Annual Report, 1922.
52. James Sirois, *Afloat in Time: Growing Up on the Rafts of a Gyppo Logger in the Coastal Canyons of British Columbia, 1930-1950* (Hagensborg: Skookum Press, 1998), 53-55; Drushka, *Working in the Woods*, 78.
53. Bonney, Prince Rupert Forest District, Management Annual Report, 1924.
54. B.C., *Report of the Forest Branch of the Department of Lands for the Year Ended Dec. 31, 1925* (Victoria: King's Printer, 1926), 11-12; B.C., *Report of the Forest Branch of the Department of Lands for the Year Ended Dec. 31, 1927* (Victoria: King's Printer, 1928), 10.
55. B.C., *Report of the Forest Branch of the Department of Lands for the Year Ending Dec. 31, 1922* (Victoria: King's Printer, 1923), 48; B.C., *Report of the Forest Branch of the Department of Lands for the Year Ending Dec. 31, 1930* (Victoria: King's Printer, 1931), 38, 43.
56. B.C., *Report of the Forest Branch of the Department of Lands for the Year Ending Dec. 31, 1929* (Victoria: King's Printer, 1930), 16; "Fire at Prince Rupert Spruce Mills," *BCL* 9 (Dec. 1925), 25.
57. "Northern Timbermen Meet," *BCL* 10 (Feb. 1926), 28; *BCL* 10 (Feb. 1926), 34; *BCL* 10 (May 1926), 32.
58. *BCL* 10 (Mar. 1926), 37; *BCL* 10 (May 1926), 32; *BCL* 10 (July 1926), 34; PRFD, AMR, 1926, GR1441, Reel B9899, File 027637, BCA.
59. PRFD, AMR, 1926.
60. Mould, *Stump Farms*, 67, 135; Nadine Assante, *The History of Terrace* (Terrace: Totem Press, 1972), 23; *BCL* 11 (Apr. 1927), 21; *BCL* 14 (June 1929), 34.
61. "Prince Rupert District," (Oct. 1927), 32; PRFD, AMR, 1928, GR1441, Reel B9899, File 027637, BCA.
62. B.C., *Report of the Forest Branch of the Department of Lands for the Year Ending Dec. 31, 1927* (Victoria: King's Printer, 1928), 5, 8; *WL* 24 (Mar. 1927), 34; *BCL* 11 (May 1927), 32; *BCL* 11 (June 1927), 34; *BCL* 11 (Sept. 1927), 32; *BCL* 11 (Oct. 1927), 32; *BCL* 12 (Apr. 1928), 30; *BCL* 11 (Nov. 1927), 32.
63. *BCL* 11 (June 1927), 34; "B.C. Sells Big Pulp Tract," *BCL* 24 (Jan. 1927), 16; *BCL* 11 (Feb. 1927), 34.
64. F. Sigmund to Minister of Lands, 25 Sept. 1928; R.C. McKenzie to Minister of Lands, 5 Jan. 1929, GR1441, Reel B4366, File 04080, BCA.
65. *BCL* 12 (Apr. 1928), 30; *BCL* 12 (July 1928), 30; *BCL* 12 (Aug. 1928), 32; G.A. Hunter, "The Prince Rupert District in 1928," *BCL* 13 (Mar. 1929), 26.
66. *BCL* 12 (May 1928), 28; PRFD, AMR, 1928.
67. *BCL* 12 (June 1928), 34; *BCL* 12 (July 1928), 30; *BCL* 12 (Sept. 1928), 30; *BCL* 13 (May 1929), 31; *BCL* 13 (Aug. 1929), 22; *BCL* 13 (Dec. 1929), 19.
68. *BCL* 12 (Aug. 1928), 32; *BCL* 12 (Sept. 1928), 30; *BCL* 13 (Jan. 1929), 34; *BCL* 13 (Feb. 1929), 30; *BCL* 13 (Apr. 1929), 30; *BCL* 13 (May 1929), 31; *BCL* 13 (July 1929), 29; PRFD, AMR, 1929, GR1441, Reel B9899, File 027637, BCA.

69. PRFD, AMR, 1929; *BCL* 13 (Apr. 1929), 30; *BCL* 13 (May 1929), 31; *BCL* 13 (Nov. 1929), 20; R.E. Allen to Chief Forester, 26 Aug. 1929, E.E. Gregg to Chief Forester, 31 Oct. 1929, GR1441, Reel B 4351, File 0772, BCA.
70. PRFD, AMR, 1929.
71. Ibid.; *BCL* 13 (Jan. 1929), 34; *BCL* 13 (Apr. 1929), 30; *BCL* 13 (June 1929), 30; *BCL* 13 (Nov. 1929), 20; *BCL* 13 (Dec. 1929), 19.

Chapter 4: From Slump to Boom

1. MacKay, *Empire of Wood,* 111; PRFD, AMR, 1930, 1-5, GR1441, Reel B9899, File 027637, BCA; *BCL* 14 (Jan. 1930), 18.
2. PRFD, AMR, 1930, 6; *BCL* 14 (Oct. 1930), 27; *BCL* 14 (Jan. 1930), 18; *BCL* 14 (Sept. 1930), 24.
3. PRFD, AMR, 1930, 3-4; *BCL* 14 (May 1930), 20; *BCL* 14 (July 1930), 26; *BCL* 14 (Aug. 1930), 26.
4. B.C. *Report of the Forest Branch of the Department of Lands for the Year Ending Dec. 31, 1931* (Victoria: King's Printer 1932), 5; PRFD, AMR, 1931, 1-2; GR1441, Reel B9899, File 027637, BCA.
5. PRFD, AMR, 1931, 3-4; A.B. Hopkinson, "A Report on a Visit to the Queen Charlotte Islands," (unpublished report, B.C. Ministry of Forests Library, 1931).
6. PRFD, AMR, 1931, 4-5; *BCL* 15 (Aug. 1931), 23; *BCL* 14 (Nov. 1930), 23; *BCL* 16 (Apr. 1932), 25; Luckhardt, "Prince Rupert," 315-16; PRFD, AMR, 3, GR1441, Reel B9899, File 027637, BCA.
7. PRFD, AMR, 1931, 5-8.
8. PRFD, AMR, 1932, 1-3, 12.
9. Ibid., 4-5; Dianne Newell, "Dispersal and Concentration: The Slowly Changing Spatial Pattern of the British Columbia Salmon Canning Industry," *Journal of Historical Geography* 14, 1 (1988), 22-36.
10. PRFD, AMR, 1932, 5-6.
11. Ibid., 6-9.
12. Ibid., 10-11.
13. PRFD, AMR, 1933, 12; GR1441, Reel B9899, File 027637, BCA; MacKay *Empire of Wood,* 113; Rajala, "There Was No Place," 14.
14. PRFD, AMR, 1931, 3-4, 11; Ramsay, *Rain People,* 152.
15. PRFD, AMR, 1931, 2-3.
16. Ibid., 11.
17. Ibid., 14.
18. Ibid., 8-9.
19. Ibid., 10-11.
20. PRFD, AMR, 1934, 1-4, 13-14, GR1441, Reel B9899, File 027637.
21. Ibid., 3.
22. Ibid., 4-6.
23. Ibid., 6-10.
24. PRFD, AMR, 1935, 1, 4-5, GR1441, Reel B9899, File 027637, BCA.
25. PRFD, AMR, 1936, 5, GR1441, Reel B9899, File 027637, BCA.
26. PRFD, AMR 1935, 2; PRFD, AMR 1936, 3; PRFD, AMR 1937, 3, GR1441, Reel B9899, File 027637, BCA.
27. PRFD, AMR 1937, 3; PRFD, AMR, 1938, 3, GR1441, Reel B9899, File 027637, BCA.
28. PRFD, AMR, 1936, 4; PRFD, AMR 1937, 4; PRFD, AMR, 1938, 1-4.
29. PRFD, AMR, 1938, 1-4; PRFD, AMR, 1939, 1, 4-5, GR1441, Reel B9899, File 027637, BCA.
30. PRFD, AMR 1937, 4; PRFD, AMR, 1938, 4; PRFD, AMR, 1939, 5.

31. PRFD, AMR, 1935, 5-7; PRFD, AMR 1936, 8; PRFD, AMR, 1937, 11.

32. PRFD, AMR, 1935, 8; PRFD, AMR, 1936, 9; PRFD, AMR, 1937, 13-14; *BCL* 19 (Nov. 1935), 16; Kerby, *One Hundred Years of History*, 17; PRFD, AMR, 1938, 10.

33. PRFD, AMR, 1936, 4-5; PRFD, AMR, 1937, 5-6; PRFD, AMR, 1938, 5.

34. PRFD, AMR, 1938, 6; PRFD, AMR, 1939, 6.

35. W. Hall, "Moresby Forest: Survey and Preliminary Management Plan", (unpublished report, B.C. Forest Branch, 1937), 19, 33.

36. Ibid., 1, 33.

37. F.D. Mulholland, *The Forest Resources of British Columbia* (Victoria: King's Printer, 1937), 75-77; D.L. McMullen, "Esquimalt and Nanaimo Land Grant: Survey and Recommendations for Improved Forest Practice" (B.C. Forest Branch, unpublished report, 1937), 1, 34.

38. See Richard A. Rajala, *Clearcutting the Pacific Rain Forest: Production, Science, and Regulation* (Vancouver: UBC Press, 1998), 160-64; Jeremy Wilson, "Forest Conservation in British Columbia, 1935-1938: Reflections on a Barren Political Debate," *BC Studies* 76 (Winter 1987-88), 3-32.

39. B.C., *Report of the Forest Branch of the Department of Lands for the Year Ended Dec. 31, 1931* (Victoria: King's Printer, 1932), 17.

40. Ibid., 18.

41. F.D. Mulholland, *The Forest Resources of British Columbia* (Victoria: King's Printer, 1937), 93-95.

42. Ken Drushka, *HR: A Biography of H.R. MacMillan* (Madeira Park: Harbour Publishing, 1995), 191-203; Taylor, *Timber*, 154-55.

43. PRFD, AMR, 1939, 1-4.

44. Ibid., 11-13.

45. PRFD, AMR, 1940, 1; PRFD, AMR 1941, 1; PRFD, AMR, 1942, 1; PRFD, AMR, 1943, 1, GR1441, Reel B9899, File 027637, BCA.

46. PRFD, AMR, 1940, 4-5; C.G. Dunham, "Trucking on Queen Charlottes," *Timberman* 42 (Mar. 1941), 34-36.

47. PRFD, AMR, 1941, 3.

48. Taylor, *Timber*, 156; MacKay, *Empire of Wood*, 143; "Aero Timber Products Limited, Gets Down to Business," *BCL* 26 (July 1942), 25; PRFD, AMR, 1942, 3.

49. PRFD, AMR, 1939, 17-18.

50. PRFD, AMR, 1940, 2-4.

51. For extensive discussion of the social and technical aspect of selective logging on the lower coast during the 1930s see Rajala, *Clearcutting the Pacific Rain Forest*, 154-66.

52. PRFD, AMR, 1940, 4.

53. Ibid., 5, 13.

54. PRFD, AMR, 1941, 3; PRFD, AMR, 1942, 3.

55. PRFD, AMR, 1943, 3; PRFD, AMR, 1944, 3; "Disband Aero Timber Products," *BCL* 29 (Sept. 1945), 35.

56. R.H. Roy, "Western Canada During the Second World War," *Journal of the West* 32 (Oct. 1993), 54-61; Kerby, *One Hundred Years of History*, 17; Luckhardt, "Prince Rupert," 318-19; PRFD, AMR, 1940, 7; PRFD, AMR, 1942, 3.

57. PRFD, AMR, 1942, 3, 6; Dirk Septer, "Highway 16: Prince Rupert-Terrace, 1944-1994," *B.C. Historical News* 29 (Winter 1995/96), 24-26.

58. PRFD, AMR, 1940, 1-2; PRFD, AMR, 1943, 5; PRFD, AMR, 1944, 5.

59. PRFD, AMR, 1943, 5; PRFD, AMR, 1944, 5; PRFD, AMR, 1945, 6.

60. Leslie Main Johnson, "Aboriginal Burning for Vegetation Management in Northwest British Columbia," in *Indians, Fire, and the Land in the Pacific Northwest*, ed. Robert Boyd (Corvallis: Oregon State University Press, 1999), 238-52.

Chapter 5: "No Camp Large or Small Will Be Missed"

1. PRFD, AMR, 1934, 1937, GR1441, Reel B9899, File 027637, B.C. Ministry of Lands Records, BCA.
2. "Aero Timber Products Gets Down to Business," *BCL* 26 (July 1942), 25
3. Neufeld and Parnaby, *The IWA in Canada*, 34-40.
4. Myrtle Bergren, *Tough Timber: The Loggers of British Columbia – Their Story* (Toronto: Progress Books, 1966), 34-52; Richard Rajala, *The Legacy and the Challenge: A Century of the Forest Industry at Cowichan Lake* (Lake Cowichan: Lake Cowichan Heritage Advisory Committee, 1993), 61-62.
5. Jerry Lembke and William Tattum, *One Union in Wood: A Political History of the International Woodworkers of America* (Madeira Park: Harbour Publishing, 1984), 36-42.
6. "Hanson Fires Delegate: Inefficiency is the Excuse," *B.C. Lumber Worker*, 4 Apr. 1936, 2 [hereafter cited as *BCLW*]; "Morgan's Camp is Confronted by Many and Varied Problems," *BCLW* 4 Apr. 1936, 3; Rajala, *The Legacy and the Challenge*, 64-65.
7. PRFD, AMR, 1936, 10; "General Conditions at Carstairs Are Fair," *BCLW*, 28 Mar. 1936, 2; "Gildersleve's Camp Has Majority Union Crew," *BCLW*, 8 Aug. 1936, 2; "Scores Overbearing Tactics of Camp Owners," *BCLW*, 18 Jan. 1936, 7; "This Outfit is a Real Frontier Outpost," *BCLW*, 5 Sept. 1936, 2.
8. "Morgan's Crew is Confronted," 3; "No First Aid, No Stretcher," *BCLW*, 26 July 1936, 8; "Union is Well Established at Camp of Kelley Log Company," *BCLW*, 14 Oct. 1936, 2; "Good Food and No Highball at Allison Log," *BCLW* 30 Sept. 1936, 2.
9. "No First Aid," 8; "No Provision Made for Removal of Injured," *BCLW*, 5 Sept. 1936, 3.
10. "Logger Hurt, Denied Aid," *BCLW*, 4 Aug. 1937, 1, 8.
11. "Morgan Crew Elects New Camp, Press and Safety Committees," *BCLW*, 9 June 1937, 2; "Morgan Crew Says Overtime Pay or No Overtime Work," *BCLW*, 6 Oct. 1937, 2; "Morgan Camps Will Continue Fight for Eight-Hour Day," *BCLW*, 13 Oct. 1937, 2; "No First Aid Attendant at Morgan's Camp This Season," *BCLW*, 22 Apr. 1938, 1, 7; "Injured At Allison's, No Doctor for 82 Hours," *BCLW*, 6 Dec. 1938, 2.
12. "Bad Accommodations Aboard Prince John," *BCLW*, 30 May 1936, 3.
13. A member, "Loggers Get Rotten Deal on Steamships," *BCLW*, 14 Oct. 1936, 4.
14. "Loggers in Morgan Camps Fed Spoiled Meats," *BCLW*, 12 July 1936, 2.
15. "Morgan Grants Temporary Wage Agreement Because of Strike," *BCLW*, 30 May 1936, 2; "Workers to Demand Fifty-Cent Wage Increase," *BCLW*, 13 June 1936, 2; "Delegation to Camp Owner Secures Wage Raise," *BCLW*, 11 July 1936, 2.
16. "Watch For the Boat," *BCLW*, 29 Aug. 1936, 6; "Gas-Boat Leaves on First Cruise to Coast Camps," *BCLW*, 5 Sept. 1936, 1.
17. A. Johnson, "Organizers' Gas-Launch Ready to Cross to Queen Charlottes," *BCLW*, 16 Sept. 1936, 7; A. Johnson, "Seagoing Organizers Are Well Received on Northern Islands," *BCLW*, 23 Sept. 1936, 1.
18. "Camp Committee is Elected at Morgan's Camp," *BCLW*, 23 Sept. 1936, 2; A. Johnson, "Union's Boat Proving Big Aid in Building a Loggers' Local," *BCLW*, 30 Sept. 1936, 2; "Union is Well Established," 2.
19. A Fisherman, "Fishermen OK Loggers' 'Navy'," *BCLW*, 30 Sept. 1936, 4; "A Good Investment," *BCLW*, 23 Sept. 1936, 6.
20. Neufeld and Parnaby, *The IWA in Canada*, 51-53; Bergren, *Tough Timber*, 111.
21. Robin Fisher, *Duff Pattullo of British Columbia* (Toronto: University of Toronto Press, 1991), 310; Neufeld and Parnaby, *The IWA in Canada*, 56.
22. "Powell River Workers Join Union," *BCLW*, 2 June 1937, 1; "Paper Plant Men Seek Agreement," *BCLW*, 28 July 1937, 1; *Labour Gazette* 38 (Dec. 1938), 406.
23. "Gas-Boat Completes Successful Cruise," *BCLW*, 26 May 1937, 1; A. Johnson, "Gas-Boat Has Been Decided Asset to Central Loggers' Local Union," *BCLW*, 30 June 1937, 2; "Diesel Installed in Union-Owned Boat," *BCLW*, 14 July 1937, 1.
24. B.H. Petersen, "The Welcome We Are Given Wherever We Go Surely is Encourage-

ment to Carry On, Says Crew of Laur Wayne," *BCLW*, 22 Sept. 1927, 4; "Cumshewa Inlet Union Camp; Reports All OK Except Eats," *BCLW*, 11 Aug. 1937, 3; Al Parkin, "Labour and Timber," *BCLW*, 21 Apr. 1947, 6.

25. "Pearson OK's Blacklisting," *BCLW*, 9 Feb. 1938, 1, 4.

26. "Annual Meet Local 71 Maps Union Program," *BCLW*, 5 July 1938, 1, 7; "$168 From QCI to Strike Fund," *BCLW*, 6 Sept. 1938, 1; Bergren, *Tough Timber*, 125, 139-41.

27. Bergren, *Tough Timber*, 140; "Loggers' Navy Reports Wage Gain From Q.C.I.," *BCLW*, 2 Apr. 1940, 1; "Q.C.I. Loggers Gain Another 50 Cent Wage Inc.," *BCLW*, 13 Nov. 1940,1; "50,000 Miles – The Saga of the Loggers' Navy," *BCLW*, 16 Apr. 1941, 1.

28. "Loggers' Flagship M.V. *Annart* Left for New Permanent Post," *BCLW*, 8 Jan. 1941, 3; "50,000 Miles," 1.

29. Rajala, *The Legacy and the Challenge*, 100-101.

30. "Loggers' Flagship M.V. *Annart* Left for New Permanent Base," *BCLW*, 8 Jan. 1941, 3; "Camp Activity," *BCLW*, 7 Feb. 1941, 2.

31. "Allison Log Provoking Crew," *BCLW*, 30 Apr. 1941, 1.

32. Laura Sefton MacDowell, "The Formation of the Canadian Industrial Relations System During World War Two," in *Canadian Working Class History: Selected Readings*, eds. Laurel Sefton MacDowell and Ian Radforth (Toronto: Canadian Scholars' Press, 1992), 579; "Allison's Camp," *BCLW*, 30 Apr. 1941, 5.

33. "Q.C.I. Camps Get Wage Increase," *BCLW*, 28 Aug. 1941, 1; "Camp Activity," *BCLW*, 30 Sept 1941, 2; Nigel Morgan, "Loggers' Navy Busy in Q.C.I.," *BCLW*, 25 Oct. 1941, 1; "Q.C.I. Loggers Win First Step To Negotiations With Pacific Mills," *BCLW*, 29 Nov. 1941, 1; John Stanton, *Never Say Die! The Life and Times of a Pioneer Labour Lawyer* (Ottawa: Steel Rail Publishing, 1987), 61.

34. "Q.C.I. Crews Start Negotiations," *BCLW*, 17 Jan. 1942, 2; "IWA Officers Negotiate With Operators of Largest Q.C.I. Camps," *BCLW*, 31 Jan. 1942, 1; Harold J. Pritchett, "Industry Council Would Settle Production Problems in Q.C.I.," *BCLW*, 28 Feb. 1942, 4; Stanton, *Never Say Die*, 61.

35. "IWA Memorandum to Operators Offers Solution to Situation in Q.C.I.," *BCLW*, 14 Mar. 1942, 5; "STOP the Sabotage of Airplane Production," (IWA, District Council No. 1, 1942); Stephen Charles Gray, "Woodworkers and Legitimacy: The IWA in Canada, 1937-1957," (PhD. Diss. Thesis, Simon Fraser University, 1989), 105.

36. Sefton MacDowell, "The Formation," 578; "First Returns of QCI Strike Vote," *BCLW*, 9 May 1942, 2; "QCI Strike Vote Nearly Unanimous," *BCLW*, 23 May 1942, 1.

37. "Five IWA Demands Granted in Queen Charlotte Islands Camps," *BCLW*, 6 June 1942, 1; "IWA Proposals for Increasing Production of Airplane Spruce in QCI Logging Camps," *BCLW*, 6 June 1942, 3.

38. Ibid.

39. "Pacific Mills' New Camp Organized," *BCLW*, 29 Aug. 1942, 6; "Crew at C-42 Send in Report," *BCLW*, 19 Sept. 1942, 2; "Crew at Camp A-35 Suggests Union Contract Will Settle All Q.C.I. Grievances," *BCLW*, 19 Sept. 1942, 2; "New Hospital at Cumshewa," *BCLW*, 19 Sept. 1942, 5; Gray, "Woodworkers and Legitimacy," 106-7.

40. Stanton, *Never Say Die*, 62-3; "I.W.A.–C.I.O. Back Gov't Appeal," *BCLW*, 10 Oct. 1942, 2.

41. "Operators' Counsel Blocks Settlement of QCI Log Dispute," *BCLW*, 7 Dec. 1942, 1; "Q.C.I. Arbitration Off to Good Start," *BCLW*, 11 Jan. 1943, 2; "Q.C.I. Arbitration Again Postponed," *BCLW*, 8 Mar. 1943, 2.

42. "Aero Camp Holds Union Meeting With McCuish," *BCLW*, 22 Mar. 1943, 2; "Camp Activity," *BCLW*, 22 Mar. 1943, 2.

43. "Loggers' Navy Will Be Tied Up in April Unless Permit Granted," *BCLW*, 22 Mar. 1943, 2; "Loggers' Navy Gets Some Fuel, Needs More Gas and Diesel Oil," *BCLW*, 5 Apr. 1943, 1; "President H. Pritchett to Tour Q.C.I. Camps," *BCLW*, 19 Apr. 1943, 1; "President Reports Successful Trip," *BCLW*, 31 May 1943, 2; "Local 1-71 Celebrates Burning 'Navy' Mortgage," *BCLW*, 17 May 1943, 2.

44. Gray, "Woodworkers and Legitimacy," 109-12; Sefton MacDowell, "The Formation," 582-84.

45. Gray, "Woodworkers and Legitimacy," 114-17; "President Reports," 2; "Local 1-71 Applies for Certification in QCI Camps," *BCLW*, 31 May 1943, 2.
46. "Majority Award Recommends Union Agreement for Q.C.I.," *BCLW*, 14 June 1943, 1; Stanton, *Never Say Die*, 65-66; "Arbitration Award," *BCLW*, 14 June 1943, 7.
47. "Negotiations on QCI Award to Start July 5," *BCLW*, 28 June 1943, 1; Q.C.I. Proposals Being Studied," *BCLW*, 12 July 1943, 2; "QCI Operators Condemned For Defying Gov't Award," *BCLW*, 26 July 1943, 1.
48. "QCI Logging Companies' Refusal to Accept Award Forces Gov't Strike Vote," *BCLW*, 26 July 1943, 2; "Queen Charlotte Islands," *BCLW*, 9 Aug. 1943, 5; "Aero Timber Will Negotiate," *BCLW*, 26 July 1943, 6.
49. *BCLW*, 22 Feb. 1943, 2; "Morgan's Camp Committee Questions Mr. Palmer," *BCLW*, 17 May 1943, 6; "Need Refrigeration," *BCLW*, 31 May 1943, 6; "Refrigeration Needed on Boats Carrying Foodstuffs," *BCLW*, 14 June 1943, 2; "Spoiled Meat," *BCLW*, 20 Sept. 1943, 5; "Camp Activity," *BCLW*, 26 July 1943, 2; "I.W.A. Urges Full Overtime Coverage for Woodworkers," *BCLW*, 26 July 1943, 8.
50. "Curry Proposals for QCI Settlement," *BCLW*, 6 Sept. 1943, 2; "Operators Refuse to Accept Gov't Proposals in QCI Dispute," *BCLW*, 6 Sept. 1943, 1; Gray, "Woodworkers and Legitimacy," 126-27.
51. "District Council Aids QCI," *BCLW*, 6 Sept. 1943, 1; "I.W.A. Restless Over Refusal of Operators to Sign Agreement," *BCLW*, 20 Sept. 1943, 1; "September 27 Named QCI Support Day," *BCLW*, 20 Sept. 1943, 1; "QCI Strike Deadline is Postponed to Oct. 7," *BCLW*, 4 Oct. 1943, 1; "Anti-Labour Policy of QCI Operators is Aid to Hitler," *BCLW*, 18 Oct. 1943, 1.
52. "Anti Labour Policy," 1; Bergren, *Tough Timber*, 220.
53. Gray, "Woodworkers and Legitimacy," 132-33; Stanton, *Never Say Die*, 67; "Labor Dept to Certify Aero Management Will Negotiate," *BCLW*, 18 Oct. 1943, 3.
54. "QCI Operators Sign Union Agreement, Strike Ends," *BCLW*, 1 Nov. 1943, 1.
55. "Lake Logging Accepts Harmen Award, Signs With Local 1-80,"*BCLW*, 1 Nov. 1943, 1; "Local 1-71 Wins Again, Aero Signs Union Contract," *BCLW*, 15 Nov. 1943, 1; "Negotiations Continue for IWA Master Contract," *BCLW*, 15 Nov. 1943, 1; "10,000 Woodworkers Under Contract," *BCLW*, 20 Dec. 1943, 7; Neufeld and Parnaby, *The IWA in Canada*, 75.
56. MacDowell, "The Formation," 589-90; Craig Heron, *The Canadian Labour Movement: A Short History* (Toronto: James Lorimer and Company, 1996), 72; Gray, "Woodworkers and Legitimacy," 138; "B.C. Adopts New Federal Labour Code," *BCLW*, 6 Mar. 1943, 1.
57. "Local 1-71 Signs Best Agreement to Date," *BCLW*, 7 Feb. 1944, 4; Gray, "Woodworkers and Legitimacy," 140.
58. "Camp Activity," *BCLW*, 29 Nov. 1943, 6; "Camp Activity," *BCLW*, 7 Feb. 1944, 4; "Camp Activity," *BCLW*, 26 June 1943, 4; "Camp Activity," *BCLW*, 24 July 1944, 4.
59. "Camp Activity," *BCLW*, 7 Aug. 1944, 4; "Camp Activity," *BCLW*, 21 Feb. 1944, 4; "Camp Activity," *BCLW*, 12 June 1944, 4; "Camp Activity," *BCLW*, 4 Sept. 1944, 4.
60. "Local 1-71 Sells 'Laur Wayne,'" *BCLW*, 3 Oct. 1944, 2; "McCuish Sees Poor Conditions on Trip," *BCLW*, 17 Oct. 1944, 4; "Interior Operation Offers Ten Cents, Violates Contract," *BCLW*, 30 June 1947, 6; "Former MP Hanson, Refuses to Bargain With Local 469," *BCLW*, 30 June 1947, 6; "Outstanding Advances Are Gained in Both Coast and Northern Agreements," *BCLW*, 14 July 1947, 1; Neufeld and Parnaby, *The IWA in Canada*, 125-32.

Chapter 6: "Era of Error"

1. PRFD, AMR, 1946, 1; PRFD, AMR, 1947, 1; PRFD, AMR, 1948, 1; PRFD, AMR 1950, 5, 9, GR1441, Reel B9899, File 027637, BCA.
2. PRFD, AMR, 1950, 1, 6; "Pacific Mills Spends Half Million on Plant Modernization," *BCL* 30 (Mar. 1946), 44; "Pacific Mills to Spend Additional $550,000 on Plant and Community Works at Ocean Falls," *BCL* 31 (July 1947), 74; "Pacific Mills Reports Net Profits Increase for the Year," *BCL* 34 (July 1950), 50.
3. *BCL* 37 (June 1953), 18; "To Direct Forest Management and Wood Utilization for Pacific Mills," *BCL* 28 (May 1944), 29; "Pacific Mills Reports Net Profits Increase for Past Year," *BCL* 54 (July 1950), 50.
4. PRFD, AMR, 1950, 6; "Pacific Mills Acquires J.R. Morgan Holdings", *BCL* 30 (Dec. 1946), 65; "Northern Pulpwood Limited," *Paper People* 1 (Jan. 1947), np.
5. "Permanent Logging Camps Springing Up Throughout Queen Charlotte Islands," *BCL* 32 (July 1948), 142; *BCL* 33 (Jan. 1949), 55; "Alaska Pine and Cellulose Announces Table of Organization," *BCL* 35 (June 1951), 76.
6. PRFD, AMR, 1949, 6-7, GR1441, Reel B9399, File 027637, BCA.
7. Ibid., 7.
8. C.D. Orchard, "Forest Working Circles: An Analysis of Forest Legislation in British Columbia as it Relates to Disposal of Crown Timber, and Proposed Legislation Designed to Institute Managed Harvesting on a Basis of Perpetual Yield," Memorandum to the Honourable Minister of Lands, 27 Aug. 1942, Box 41, Premiers' Papers, GR1222, BCA.
9. Keith Reid and Don Weaver, "Aspects of the Political Economy of the B.C. Forest Industry," in *Essays in B.C. Political Economy,* eds., Paul Knox and Philip Resnick (Vancouver: New Star Books, 1974), 16-19; Rajala, *Clearcutting the Pacific Rain Forest,* 192.
10. Gordon Sloan, *Report of the Commissioner Relating to the Forest Resources of British Columbia* (Victoria: King's Printer, 1945), 143-44.
11. "Huge Cellulose Development for British Columbia," *BCL* 31 (Apr. 1947), 94.
12. "Telkwa-Smithers," *Cariboo and Northern B.C. Digest* 3 (Fall 1947), 54; "Terrace," *Cariboo and Northern B.C. Digest* 3 (Fall 1947), 54; Mike Sahonovitch, "Skimmed Milk and 100 Years of Progress," *Cariboo and Northern B.C. Digest* 3 (Winter 1947), 112.
13. "First Forest Management Licence in B.C. Issued to Cellulose Firm," *BCL* 32 (May 1948), 57; "Cellulose Plant Development Will Hasten Growth of Northern B.C. Seaport," *BCL* 32 (Sept. 1948), 95; "$25 Million Cellulose Plant Aids Skeena Valley Development," *Cariboo and Northwest Digest* 5 (Oct. 1949), 19, 29.
14. Arthur C. Downs, "From Trees to Textiles,' *Cariboo and Northwest Digest* 6 (May-June 1950), 14, 16; Luckhardt, "Prince Rupert," 321.
15. Downs, "$25 Million Dollar," 19, 24; PRFD, AMR, 1949, 1.
16. "B.C.'s First Big Log Drive," *BCL* 34 (May 1950), 40-41, 81; "First B.C. Coast Log Drive Under Way on Skeena," *BCL* 34 (Sept. 1950, 43, 58, 60; PRFD, AMR, 1950, 1; PRFD, AMR, 1951, 17; GR1441, Reel B6899, File 027637, BCA.
17. "Columbia Cellulose Plant Dedicated," *BCL* 35 (June 1951), 41, 62; PRFD, AMR, 1950, 11.
18. PRFD, AMR, 1951, 16-17; PRFD, AMR, 1952, 17; "Columbia Cellulose Plant Dedicated," 41; "Spar Trees No Longer Stay in One Place," *Truck Logger* (Oct. 1954), 26, 30 [hereafter cited as *TL*].
19. Arthur Downs, "Terrace: The Town With a Future," *Northwest Digest* 11 (Jan./Feb. 1955), 10, 17, 19; Kerby, *One Hundred Years of History,* 19.
20. PRFD, AMR, 1953, 5-6; PRFD, AMR, 1954, 5; PRFD, AMR, 1955, 6, GR1441, Reel B9899, File 027637, BCA.
21. Bev Christensen, *Too Good To Be True: Alcan's Kemano Completion Project* (Vancouver: Talon Books, 1995), 79-101; Beck, *Three Towns,* 43-64; "Alcan Underwrites Logging Operation," *TL* (Dec. 1952), 12; PRFD, AMR, 1956, GR1441, Reel B9899, File 027637, BCA.
22. PRFD, AMR, 1953, 1; PRFD, AMR, 1954, 21, 25; Taylor, *Timber,* 167; Drushka, *Working in the Woods,* 253; "CZ Records Progress in Integration," *BCL* 41 (Apr. 1957), 74.

23. "B.C.'s Forest Industries Merger Taking Shape on the Coast," *BCL* 35 (Apr. 1951), 39; "Rayonier Buys Alaska Pine Stock," *TL* (Dec. 1954), 11; Taylor, *Timber*, 125-26; McKay, *Empire of Wood*, 88, 214-30.

24. Van Perry, "Rising Pressure in Forestry Situation: Enquiry Due," *BCL* (Dec. 1954), 50-53; Gordon Sloan, *Report of the Commissioner Relating to the Forest Resources of British Columbia,* Vol. 1 (Victoria: Queen's Printer, 1956), 104, 112; PRFD, AMR, 1954, 2; PRFD, AMR, 1955, 2, GR1441, Reel B9899, File 027637, BCA.

25. "Some FMLs to be Held Back," *BCL* 39 (Mar. 1955), 12; David J. Mitchell, *W.A.C. Bennett and the Rise of British Columbia* (Vancouver: Douglas & McIntyre, 1983), 218-50.

26. Crown Zellerbach Canada Ltd., Canadian Western Lumber Company Ltd., Elk Falls Ltd., "A Brief to the Royal Commission of Inquiry, Forest Resources of British Columbia," Dec. 1955, B.C. Legislative Library; Alaska Pine and Cellulose Ltd., "Submission to the Royal Commission on Forestry," Feb. 1956, B.C. Legislative Library; MacMillan Bloedel Ltd., "Brief to be submitted to the Royal Commission on Forestry," Nov. 1955, B.C. Legislative Library.

27. Sloan, *Report of the Commissioner*, Vol. 1, 44, 89-91, Vol. 2, 765; Sloan Report Advocates Permanent Advisory Committee," *BCL* 41 (Sept. 1957), 84-85, 93; "Substitute FML Tenure," *BCL* 42 (Apr. 1958), 86; "TFLs in Budget Speech," *BCL* 43 (Mar. 1959), 37.

28. PRFD, AMR, 1956, 1, GR1441, Reel B9899, File 027637, BCA.

29. H. Sinclair to District Forester, 19 Feb. 1948, GR1441, Reel B4366, File 04080, BCA.

30. H. Sinclair to E.T. Kenney, 21 Mar. 1950, L.G. Taft to District Forester, 2 May 1950, GR1441, Reel B4366, File 04080, BCA.

31. E.T. Kenney to H. Sinclair, 3 Apr. 1950; M.W. Gormely to C.D. Orchard, 5 May 1950, GR1441, Reel B4366, File 04080, BCA.

32. L.G. Taft to M.W. Gormely, 22 May 1950, GR1441, Reel B4366, File 04080, BCA.

33. J.V. Boys to E.T. Kenney, 30 Oct. 1950; Kenney to Boys, 10 Nov. 1950, GR1441, Reel B4366, File 04080, BCA.

34. H. Sinclair to E.T. Kenney, 22 Mar. 1951; Kenney to Sinclair, 19 Apr. 1951; "Petition to the British Columbia Forest Department," 10 Dec. 1952, GR1441, Reel B4366, File 04080, BCA.

35. J. Mould to District Forester, Prince Rupert, 30 Dec. 1952; P. Young to Chief Forester, 5 Jan. 1953; S.E. Marling to Prince Rupert, 24 Jan. 1953, GR1441, Reel B4366, File 04080, BCA.

36. F. Howard to P. Young, 17 Nov. 1954; Kitwancool Timber Sale Contractors to P. Young, 25 Nov. 1954; Young to Howard, 23 Nov. 1954; Young to W.S. Douse, 3 Dec. 1954, GR1441, Reel B4366, File 04080, BCA.

37. "To the Government of the Province of British Columbia – Forest Branch", 10 Dec. 1955, GR1441, Reel B4366, File 04080, BCA.

38. PRFD, AMR, 1955, 24-25.

39. MacDonald, "The Marginalization of the Tsimshian Cultural Ecology," 210-16; Charles R. Menzies and Caroline F. Butler, "Working in the Woods: Tsimshian Resource Workers and the Forest Industry of British Columbia," *American Indian Quarterly* 25 (Summer 2001), 409-30.

40. PRFD, AMR, 1961, 3, GR1441, Reel B9899. File 027637, BCA; "Indian Land Controversy," *TL* 47 (Sept. 1963), 8; Daniel Raunet, *Without Surrender, Without Consent: A History of the Nishga Land Claims* (Vancouver: Douglas & McIntyre, 1996), 144-46, 181; Paul Tennant, *Aboriginal Peoples and Politics: The Indian Land Question in British Columbia, 1849-1989* (Vancouver: UBC Press, 1990), 122-23.

41. Arthur Downs, "Prince Rupert: Gateway to the Pacific Northwest," *Northwest Digest* 10 (Nov./Dec. 1954), 21; PRFD, AMR, 1956, 29-31.

42. PRFD, AMR, 1958, 35-37; PRFD, AMR, 1959, 32-33, 37; PRFD, AMR, 1960, 28, GR1441, Reel B9899, File 027637, BCA; "New Operations Commenced," *Timber of Canada* 20 (July 1959), 54; "Columbia Cellulose Commences Nass River Logging," *TL* 15 (Sept. 1959), 24-25; "River Drive Going Strong," *Forest Industries* 91 (May 1964), 74-76.

43. Raunet, *Without Surrender*, 146-51, 182-83.
44. PRFD, AMR,1957, 32-33; PRFD, AMF., 1958, 38; PRFD, AMR, 1959, 35; PRFD, AMR, 1960, 29.
45. PRFD, AMR, 1961, 33; PRFD, AMR, 1962, 35, GR1441, Reel B9899, File 027637, BCA.
46. W.C. Gardiner to P. Young, 29 Dec. 1955, GR1441, Reel B4366, File 04080, BCA.
47. PRFD, AMR, 1958, 11, 20; PRFD, AMR, 1960, 3.
48. PRFD, AMR, 1957, 10; PRFD, AMR, 1958, 35; MacKay, *Empire of Wood*, 255-56.
49. PRFD, AMR, 1955, 17; PRFD, AMR, 1958, 8; PRFD, AMR, 1959, 1; PRFD, AMR, 1961, 2.
50. Clifford R. Kopas, "Bella Coola," *Cariboo and Northern B.C. Digest* 3 (Winter 1947), 44; "End of an Era for Northcop Logging Company," *Westcoast Logger* 1 (May 1990), 23-25.
51. Cliff Kopas, "Canada's Third Outlet to the Pacific," *Northwest Digest* 11 (June-July-Aug. 1955), 31; A. Sahonovitch, "Forest Management and Utilization – 1955 Style," *Northwest Digest* 11 (June/July/Aug. 1955), 38-41.
52. Kopas, "Canada's Third Outlet," 45.
53. S. Saugstad to P. Young, 17 Mar. 1955, GR1441, Reel B4366, File 04080, BCA.
54. P. Young to Northcop Logging Company, 5 Apr. 1956, GR1441, Reel B4366, File 04080, BCA.
55. S. Saugstad to District Forester, 18 May 1956; E. Knight to Northcop Logging Company, 6 June 1956, GR1441, Reel B4366, File 04080, BCA.
56. S. Saugstad to R. Williston, 1 Dec. 1957, GR1441, Reel B4366, File 04080, BCA.
57. P. Young to Chief Forester, 31 Dec. 1957; R. Williston to S. Saugstad, 28 Mar. 1958, GR1441, Reel B4366, File 04080, BCA; PRFD, AMR, 1960, 3; PRFD, AMR, 1962, 25; "End of an Era," 23.
58. PRFD, AMR, 1960, 1-2; PRFD, AMR, 1961, 20-27, 43; PRFD, AMR, 1962, 31.
59. PRFD, AMR, 1960, 2; PRFD, AMR,1962, 1, 4, 16-17; PRFD, AMR, 1955, 1, 14.
60. PRFD, AMR, 1959, 18; PRFD, AMR, 1960, 13-14; PRFD, AMR, 1962, 14.
61. PRFD, AMR, 1955, 9; PRFD, AMR, 1961, 9-10, 14; Ken Drushka, *Tie Hackers to Timber Harvesters: The History of Logging in British Columbia's Interior* (Madeira Park: Harbour Publishing, 1998), 133-35.
62. Rajala, *Clearcutting the Pacific Rain Forest*, 21C-13; PRFD, AMR, 1959, 29.
63. Rajala, *Clearcutting the Pacific Rain Forest*, 213-14.
64. PRFD, AMR, 1959, 35-36; PRFD, AMR, 1960, 22-23.
65. PRFD, AMR, 1961, 19-21, 33; PRFD, AMR, 1962, 16, 20.
66. British Columbia, *The Prince Rupert – Smithers Bulletin Area* (Victoria: Queen's Printer, 1959).
67. "The Skeena Valley," *BCL* 47 (Oct. 1963), 52-54.
68. "The New Frontier," *TL* (Oct. 1969), 7-12; Kerby, *One Hundred Years*, 19; PRFD, AMR, 1962, 2-3; "The Skeena Valley," 54.
69. Pat Carney, "The Kitimat Pulp and Paper Puzzle," *TL* 18 (Dec. 1962), 3; "Kitimat Logging Planned," *Crown Zellerbach News* 6 (Jan. 1963), 1; McKay, *Empire of Wood*, 259; "Skeena Timbermen Worried," *BCL* 48 (Feb. 1964), 40.
70. "CZ at Kitimat," *Woodlands' Review* (1963), 83; "First Tow From Kitimat," *Crown Zellerbach News* 7 (Aug. 1963), 1; "Logging Starts at Kitimat," *BCL* 48 (Sept. 1964), 34; "Interest in Kitimat Potential," *BCL* 47 (Sept. 1963), 60; Eileen Williston and Betty Keller, *Forests, Power and Policy: The Legacy of Ray Williston* (Prince George: Caitlin Press, 1997), 153.
71. "Scramble for Timber Harvesting Rights," *BCL* 48 (Mar. 1964), 44-47; Williston and Keller, *Forests, Power and Policy*, 151-54.
72. Williston and Keller, *Forests, Power and Policy*, 154-55.
73. "Extend Terrace S.Y.U.," *TL* 20 (Mar. 1964), 81; "B.C. Forest Giants Stake Claims to Timber," *Canadian Forest Industries* 84 (May 1964), 81 [hereafter *CFI*].
74. "Harvesting Rights Key to Pulp Mill Development," *BCL* 48 (Sept. 1964), 16-22; "Compromise Plan: Three-Way Split to Provice for Major B.C. Pulp Mills," *TL* 20 (Sept. 1964), 16-18.
75. Pat Carney, "Timber Rights: A Solomon's Judgement," *CFI* 84 (Sept. 1964), 84-85; Pat

Carney, "Let the Chips Fall," *CFI* 84 (Oct. 1964), 49; "MB & PR Okays Kitimat Mill," *CFI* 84 (Nov. 1964), 19.

76. Pat Carney, "Let the Chips Fall," *CFI* 85 (Jan. 1965), 58.

77. "Not Enough Timber, So MB & PR Dumps Kitimat," *CFI* 85 (May 1965), 89; MacKay, *Empire of Wood,* 259-64; R.G. Williston, "Progress Toward Maximum Yield in B.C.'s Forests," *PPMC* 67 (Jan. 1966), 5-7.

78. Williston and Keller, *Forests, Power and Policy,* 157-58.

79. "Eurocan Pulp and Paper Awarded Kitimat TFL," *Hiballer* 16 (Oct. 1965), 42, 50; "Eurocan Beats C-Z for Kitimat Timber," *CFI* 85 (Nov. 1965), 49.

80. "Eurocan Pulp Mill for Kitimat," *TL* 21 (Nov. 1965), 34, 37.

81. "Eurocan Buys Site From Alcan at Kitimat," *TL* 23 (June 1967), 12; "Oct. 1, 1970 Date for Eurocan Mill Start Up," *CFI* 88 (May 1968), 15; "Eurocan Breaks Ground for $100 Million Mill," *CFI* 88 (July 1968), 10; "Eurocan Lumber Output to Start Next Summer," *CFI* 88 (Aug. 1968), 17; Williston and Keller, *Forests, Power and Policy,* 158-59.

82. David Gray, "Kitimat," *BCL* 52 (Oct. 1958), 28-33.

83. "Capacity Boost Planned for Eurocan Pulp, Paper and Lumber Mill," *PPMC* 70 (6 June 1969), 13; "Eurocan Pulp and Paper to Expand Mill by One-Third," *PPMC* 70 (3 Oct. 1969), 19.

84. "Woodmill Section of $145 Million Eurocan Compler Now Operating," *Hiballer* 21 (Mar. 1970), 26-28; "Eurocan Starts Production," *Hiballer* 21 (Oct. 1970), 28-29; "By Lake, Land and Sea," *Hiballer* 21 (Oct. 1970), 44-46; "Logs Hauled on Continuous Basis During Four Months," *Forest Industries* 97 (Nov. 1970), 51; "Giants at Work," *BCL* 55 (Jan. 1971), 10-11.

85. "Interest in Pulp Mill Sold by Ben Ginter," *CFI* 9 (Feb. 1971), 21; "May Be 1972 Before Recovery Takes Hold, Says Eurocan's Dunn," *Canadian Pulp and Paper Industry* 24 (Aug. 1971), 29-31 [hereafter *CPPI*]; "100-Ton Bundles Trim Costs on Interior-to-Coast Log Haul," *CFI* 92 (Aug. 1972), 34-38; Drushka, *Tie Hackers to Timber Harvesters,* 191.

86. "100-Ton Bundles," 38; Rajala, *Clearcutting the Pacific Rain Forest,* 44-49.

87. Miles Overend, "Babine Forest Products: A Piece of the Action for Native Indians," *CFI* 95 (Oct. 1975), 34-37; Miles Overend, "Babine Opens 110-Million-BF Mill," *CFI* 95 (Oct. 1975), 45-47.

88. "West Fraser Buys Into Eurocan," *Pulp and Paper* 53 (Nov. 1979), 15; "West Fraser," *Eurocan News* 6 (Dec. 1979), 14-15; "West Fraser May Merge With Eurocan," *CPPI* 33 (Aug. 1980), 10; "West Fraser Buys 40% of Eurocan, Plans $100 Million Expansion Program," *CPPI* 34 (May 1981), 8; Drushka, *Tie Hackers to Timber Harvesters,* 218.

89. "Columbia Cellulose Announces Partnership Formed to Build New Prince Rupert Pulp Mill," *Hiballer* 15 (Nov. 1964), 34-36; "New Prince Rupert Mill," *BCL* 48 (Dec. 1964), 41; Drushka, *Tie Hackers to Timber Harvesters,* 180; Luckhardt, "Prince Rupert," 323.

90. Canadian Cellulose Company Ltd., "Submission to the Royal Commission on Forest Resources," 1975, 4-5; Jeremy Wilson, *Talk and Log: Wilderness Politics in British Columbia* (Vancouver: UBC Press, 1998), 88; G.L. Ainscough, "The British Columbia Forest Land Tenure System," in *Timber Policy Issues in British Columbia,* eds. William McKillop and Walter J. Mead (Vancouver: UBC Press, 1976), 51.

91. Canadian Cellulose Company Ltd., "Submission," 5; "$80 Million Skeena Kraft Claims Largest Since Pulp Mill," *Hiballer* 18 (Sept. 1967), 41.

92. "Columbia Cellulose Sponsor Housing Program," *PPMC* 67 (Feb. 1966), 12; "First Home Opened at New Mill," *TL* 22 (Feb. 1966), 1; "Forest Development Opening Up Wonders of Northwest," *Hiballer* 18 (May 1966), 41-44; "Skeena Kraft – New Giant on Canada's Pacific Coast," *PPMC* 68 (July 1967), 94-103; "Twinriver Timber Handles Log Supply," *TL* 23 (July 1967), 30-33; "Planetary Axles Bear Brunt of 125-Ton B.C. Log Trucks," *CFI* 89 (Nov. 1969), 53-54; "Jet Boats on the Nass River," *BCL* 57 (Mar. 1973), 16-18.

93. Canadian Cellulose Company Ltd., "Submission," 6; "Skeena Kraft Buys Timber at Hazelton," *TL* 23 (June 1967), 12; "Columbia Cellulose Cuts Projects, Staff at Research Division," *PPMC* 69 (2 Feb. 1968), 13; "Pulp Owners Change," *TL* 24 (Aug. 1968), 29; "Celanese Sells Interest in Colcel," *PPMC* 70 (21 Mar. 1969), 14; "Colcel Profit Finally,"

PPMC 70 (18 Apr. 1969), 14; "Colcel Buys Pohle, Plans Forest Complex," *CFI* 89 (Nov. 1969), 23; "Colcel Stays in Black During First Quarter, But Still Curtails Woods Operations," *PPMC* 71 (1 May 1970), 17-18; Luckhardt, "Prince Rupert," 324.

94. Canadian Cellulose Company Ltd., "Submission," 10-11; "Colcel Curtails Wood Production," *CFI* 90 (May 1970), 21;

95. Frank Zelko, "Making Greenpeace: The Development of Direct-Action Environmentalism in British Columbia", *BC Studies* 142/143 (Summer/Autumn 2004), 197-239; "Colcel Suing Dupont Over Pulp Pipeline," *CFI* 90 (Sept. 1970), 15; "Columbia Cellulose Fined $3,000," *BCL* 54 (Oct. 1970), 4; "How Colcel Handles Chemical Process Change," *CPPI* 23 (Dec. 1970), 34.

96. Wilson, *Talk and Log*, 80; Yasmeen Quereshi, "Environmental Issues in British Columbia: An Historical-Geographical Perspective," (M.A. thesis, University of British Columbia, 1991), 110-11.

Chapter 7: Winding Down

1. Ray Williston, "Foreign Capital Investment Defined," *TL* 25 (Jan. 1969), 50.

2. Wilson, *Talk and Log*, 107; "Will Set Land Use Priorities," *BCL* 54 (Sept. 1970), 53; Robert Williams, "Ripping Off B.C.'s Forests," *Canadian Dimension* 7 (Jan./Feb. 1971), 19-22.

3. "B.C. Forest Industry Fearful of NDP Aims," *CFI* 98 (Oct. 1972), 17; Dave Barrett and William Miller, *Barrett: A Passionate Political Life* (Vancouver: Douglas & McIntyre, 1995), 58.

4. "Ocean Falls Exporting Pulp," *PPMC* 67 (Sept. 1966), 24; "CZ Studies Changes for Ocean Falls," *CFI* 85 (May 1965), 89; "CZ to Phase Out Two Ocean Mills," *CFI* 86 (Mar. 1967), 67; "Ocean Falls Revamped From the Groundwood Up," *CPPI* 21 (Jan. 1968), 24; "CZ to Modernize Ocean Falls Operations," *TL* 23 (Mar. 1967), 16; "CZ Workforce to be Cut by 150," *CPPI* 21 (Feb. 1968), 4.

5. "CZ to Close Pulp Mill at Ocean Falls," *CFI* 92 (June 1972), 20; Gunter Hogrefe, "End of an Era: Ocean Falls: 1971 to 1980," in Bruce Ramsay, *Rain People: The Story of Ocean Falls*, Second Edition (Kamloops: Wells Gray Tours, 1997), 217-24.

6. "Colcel Submits Pollution Plan," *CPPI* 24 (Feb. 1971), 10; "Colcel Completes Project at Prince Rupert Complex," *BCL* 56 (Jan. 1972), 7, 16; "Celanese Wants to Sell Columbia Cellulose," *CFI* 92 (Mar. 1972), 11; Kerby, *One Hundred Years*, 21; G.W. Taylor, "A Tale of Three Companies," *Hiballer* 23 (Mar. 1972), 48-51.

7. Wilson, *Talk and Log*, 127-28; "B.C. Takeover of Colcel is Under Way," *CPPI* 26 (Apr. 1973), 18; "B.C. Acquires Ailing Colcel Operations," *CFI* 93 (May 1973), 11; Ted Stevens, "Government and the Forest Industry," *B.C. Logging News* 1 (May 1974), 15-18 [hereafter *BCLN*].

8. "Canadian Cellulose Company," *Forestalk* 1 (Autumn 1973), 3-7.

9. The Big Chip Hastle," *BCL* 57 (Aug. 1973), 14; Drushka, *Tie Hackers to Timber Harvesters*, 182; "Canadian Cellulose Company," 6; Taylor, "A Tale of Three Companies," 52.

10. "B.C. to Take Over Ocean Falls Site," *CFI* 92 (Oct. 1972), 18; "B.C. Pays $1 Million for CZ Ocean Falls Operation," *CFI* 93 (May 1973), 27; "Government Takeover Brings New Life to Ocean Falls," *Forestalk* 1 (Summer 1973), 4-6; Hogrefe, "End of an Era," 220-24.

11. "Ocean Falls Reports Loss," *CPPI* 27 (June 1974), 11; "Report Recommends Growth at Ocean Falls," *CFI* 94 (May 1974), 15; "Ocean Falls B.C. Cel Report '74 Profits," *CFI* 95 (Aug. 1975), 10; "Major Expansion of British Columbia's Forest Industry," *CFI* (May 1975), 3; Williston and Keller, *Forests, Power and Policy*, 268.

12. Diane Newell, *Tangled Webs of History: Indians and the Law in Canada's Pacific Coast Fisheries* (Toronto: University of Toronto Press, 1993), 126-28, 148-60; W.C. Cheston, R.B. Allison, V.A Holm and W.G. Swanson, *The Bella Coola Regional Study*, Vol. 1 (Victoria: B.C. Forest Service, Special Studies Division, 1975), 16-17.

13. Cheston, et.al., *The Bella Coola Regional Study*, 22, 48-50.
14. Cheston, et.al., *The Bella Coola Regional Study*, 25, 80, 96; "Forest Service to Study Possible Road," *BCL* 54 (July 1971), 29.
15. Cheston, et.al., *The Bella Coola Regional Study*, 25; Crown Zellerbach Canada Ltd. "Submission to the Royal Commission on Forest Resources," Nov. 1975, 18-20.
16. Cheston, et.al., *The Bella Coola Study*, 114.
17. Cheston, et. al., *The Bella Coola Study*, 114-15.
18. "1974 Earnings of $29 Million Fuels Inborn Optimism of Cancel's R. Gross," *CPPI* 28 (Mar. 1975), 17-18; "Guidelines for the Coast Forest Industry," *TL* 28 (Dec. 1972), 28-30; "B.C. Loggers Protest Environmental Guidelines," *CFI* 93 (June 1973), 28-29; "B.C. Logging Industry Hamstrung by Gov't Red Tape," *BCLN* 1 (Nov. 1974), 21-24.
19. "Stumpage Rates Reduced," *BCLN* (Sept. 1974), 44; Wilson, *Talk and Log*, 128.
20. *CFI* 95 (June 1975), 2; Evelyn Pinkerton, "Taking the Minister to Court: Changes in Public Opinion About Forest Management and Their Expression in Haida Land Claims," *BC Studies* 57 (Spring 1983), 74-75.
21. Jean Sorensen, "Environmental Squabble Dies Natural Death," *BCL* 59 (July 1975), 50-51; Pinkerton, "Taking the Minister to Court," 76-78; "Williams Bypasses Forest Service," *BCLN* 6 (Feb./Mar. 1975), 52-53.
22. Rayonier Canada Ltd., "Submission to the Royal Commission on Forest Resources," Nov. 1975, 16-17, B.C. Ministry of Forests Library; "Peter Pearse: Chief Shrink to the Forest Industry," *BCLN* 7 (Jan. 1976), 25-27.
23. "Proposal to be Submitted to the Forest Resources Commission," Skidegate Band Council, 11 Sept. 1975, 2-4, B.C. Ministry of Forests Library.
24. Ibid., 5-6.
25. B.C. Royal Commission on Forest Resources, *Proceedings*, Sept. 1975, Vol. 12, 1953-55.
26. Ibid., 1968-75.
27. B.C. Royal Commission on Forest Resources, *Proceedings*, Vol. 15, 16 Sept. 1975, 2586.
28. Nass Valley Communities Association, "Logging and Small Communities, Report to the Royal Commission on Forest Resources," 1975, 1-7, B.C. Ministry of Forests Library.
29. Nass Valley Communities Association, "Logging and Small Communities," 2-9.
30. Kispiox Valley Community Association, "Submission to the Royal Commission on Forest Resources," Dec. 1975, 7-8; B.C. Ministry of Forests Library; Nass Valley Communities Association, "Logging and Small Communities," 5-6.
31. Kispiox Valley Community Association, "Submission," 1-12.
32. Lorne J. Kravic and Gary Brian Nixon, *The 1200 Days: A Shattered Dream, Dave Barrett and the NDP in B.C., 1972-75* (Coquitlam: Kaen Publishers, 1978), 12; "New Forests Boss," *BCLN* 7 (Jan. 1976), 30.
33. "Logging Contractors Told They'll Have to Move Out," *BCLN* 7 (May 1976), 296-97; Kerby, *One Hundred Years*, 22; *CFI* 96 (Jan. 1976), 5.
34. Peter H. Pearse, *Timber Rights and Forest Policy in British Columbia: Report of the Royal Commission on Forest Resources*, Vol. 1 (Victoria: Queen's Printer, 1976), 2-7, 22; Wilson, *Talk and Log*, 152-53.
35. Wilson, *Talk and Log*, 24; Pearse, *Timber Rights*, 30-34, 78-81, 92-93.
36. Pearse, *Timber Rights*, 94-96; Wilson, *Talk and Log*, 159; Richard Schwindt, "The Pearse Commission and the Industrial Organization of the British Columbia Forest Industry," *BC Studies* 41 (1979), 3-35.
37. "A Few Highlights," *BCL* 62 (June 1978), 32; "Forests Minister Outlines Philosophy," *Journal of Logging Management* 10 (Mar. 1979), 1883; Jean Sorensen, "A Perspective on Forest Policy: the New Legislation," *Forestalk* 2 (Summer 1979), 21-22; Patricia Marchak, *Green Gold: The Forest Industry in British Columbia* (Vancouver: UBC Press, 1983), 55-58; Wilson, *Talk and Log*, 159.
38. Marchak, *Green Gold*, 104; "Cancel Now Wholly Owned by BRIC," *Canadian Cellulose News* 13 (Aug. 1980), 1; Jean Sorensen, "Has B.C. Assured Cancel's Development," *BCL* 61 (Oct. 1977), 71-72; Larry D. Skory, "Cancel Choses Kraft," *CPPI* 30 (Nov. 1977), 12-